Beyond Technocracy

Massimiano Bucchi

Beyond Technocracy

Science, Politics and Citizens

Springer

Massimiano Bucchi
Università di Trento

Translated by
Adrian Belton

Translated from the Italian *Scegliere il mondo che vogliamo. Cittadini, politica, tecnoscienza* by Massimiano Bucchi, published by Società editrice il Mulino S.p.a., Bologna, Italy © 2006.

ISBN 978-0-387-89521-5 e-ISBN 978-0-387-89522-2
DOI 10.1007/978-0-387-89522-2
Springer Dordrecht Heidelberg London New York

Library of Congress Control Number: 2009928234

Printed on acid-free paper

Springer is part of Springer Science+Business Media (www.springer.com)

Acknowledgements

I should thank Barbara Allen, Piero Bassetti, Alessia Graziano, Bruno Latour, Federico Neresini, Giuseppe Pellegrini, Roberto Franzosi, Pierangelo Schiera, Mariachiara Tallacchini, Giuseppe Testa and Bryan Wynne for their discussions and encouragement; Marco Cavalli and Renato Mazzolini for reading and commenting on earlier versions of this manuscript; Marco Brunazzo and Mario Diani for important bibliographic suggestions on Chap. 3; Adrian Belton and Alessia Bertagnolli for their translating and editing work. Some of the arguments presented here have been discussed during seminars at the European University Institute, the Center for Genomics and Society dell'Università di Exeter and the London School of Economics. I should thank all participants for their comments and in particular Donatella Della Porta, Martin Bauer and Massimo Mazzotti.

Contents

Introduction

Science and Society: A Clash of Civilizations?

Nuclear energy, GM foods and stem cells: the more rapidly science advances, the more society seems to resist it. Not a day passes without the news media reporting protests against genetically modified foods, demonstrations and road blocks against the disposal of radioactive waste, motions lodged with the international institutions and heated controversy on embryo stem-cell research.

Issues concerning scientific research and technological innovation appear with increasing frequency on the agendas of public and political debate. The outcome in many cases is an open conflict among technical-scientific experts, policy-makers, business lobbyists and citizens, which not infrequently paralyses decision making.

The proliferation of conflicts on techno-scientific issues raises a series of questions. Are we witnessing a radical clash between science and society? How have we come to this pass? Are our institutions – from political to scientific – capable of meeting the challenges raised by research and technological innovation? Are citizens sufficiently well informed to discuss them? What scenarios await us in the future? What responses and strategies can help decision makers tackle these issues? In short, how can the increasing need to take decisions on highly complex technical-scientific matters be reconciled with rights to democratic participation?

This book argues that these issues and conflicts cannot be considered episodic; nor can they be glibly dismissed with epithets like "obscurantism", "anti-scientism" or "scientific illiteracy". The book puts forward two hypotheses: first that these cases are symptomatic of major – perhaps even epochal – changes in the social role of science, and generally in the production of scientific knowledge; second that such changes concern the nature itself of contemporary politics and democracy.

The first chapter considers a particularly common type of response to the emergence of issues and conflicts in technoscience: the so-called "technocratic" response. The chapter seeks to show the principal reasons why there seems to be no way out of the decision-making deadlock on technoscience and from the sterile opposition between science and society viewed as a "clash of civilizations".

The book's two central chapters analyse two key factors in understanding and tackling the challenges of technoscience in contemporary society. Specifically, the second chapter examines the significant changes in the processes of knowledge production which have led to so-called "post-academic" science. The third chapter explores the proliferation of participation by non-experts in such processes – corresponding to which is the increasingly large-scale mobilization of technical-scientific experts in the public arena.

On this basis, the last chapter urges that the illusory shortcuts offered by technocracy and frequent appeals to individual ethics should be eschewed. It describes the substantial implications that full and mature awareness of the challenges and conflicts on science entail, not only for the scientific sphere, but also for democratic politics.

A terminological specification is necessary. I shall frequently use the term *technoscience* to denote both scientific research and technological innovation. The use of this term, however, should not be taken as implying a substantial overlap between the two areas. Nor is it intended to revive debate on the relationship between science and technology, which has been already amply discussed elsewhere. Here the term refers to two phenomena in particular. First, the increasing closeness between the contexts of research and those of its application, which is considered one of the key features of the current configuration of contemporary science – so-called "post-academic" science.[1] In this sense, the term "technoscience" can be considered similar in meaning to "knowledge". It is not a mere stock of expertise but a combination of knowledge and power, or the "generalized capacity for action" that some scholars regard as typical of contemporary knowledge-based societies (Stehr 2005:24).

Second, the overlap between research and innovation – however, debatable in its substance – is a distinctive feature of the discourse on, and the public perception of, science issues. Yet, however, close or distant science and technology may be in the concrete practice of researchers (and innovators), or in the analyses of scholars, in the public arena; they are frequently treated as being one and the same.

[1] Gibbons et al. (1994); Ziman (2000); Nowotny et al. (2001).

Chapter 1
The Technocratic Response: All Power to the Experts

(With the Blessing of Citizens Provided They Are Well-Educated)

The populace was confused, yet didn't care.
Scientists peered into data and concluded
that we should all be very worried

(Bret Easton Ellis, Lunar Park)

1.1 The "Missionary" Wing of Technocracy: "Deficit" and the Public Understanding of Science

This first chapter considers a particularly common response to issues and conflicts in technoscience. I refer to the so-called "technocratic response", which is especially frequent within the scientific community, but not uncommon among policymakers and other authoritative commentators, or in some sectors of public opinion. It bases its proposals for decision making on a specific conception of the relationships among scientific experts, political decision makers, and public opinion.

Summarizing to the extreme, the technocrat conception rests on two main tenets:

(a) Public opinion and political decision makers are extremely misinformed about science and the issues raised by its development.
(b) This misinformation is fuelled by inadequate and sensationalist media coverage of technoscientific topics. This situation is exacerbated by poor training in basic science and a general disinterest – among the institutions and the cultural intelligentsia – in scientific research. Consequently, citizens and political decision makers easily fall prey to "irrational" fears which stoke their hostility and suspicion towards entire sectors of research and technological innovation (nuclear energy, GM foods, and stem cells).

Various versions of this argument have been put forward. In Italy, the tenacious hostility (indeed "obscurantist" prejudice) against science and technology has been blamed on the prevalence of philosophical-political doctrines: historicist-Crocean, Marxian, and also Catholic. A significant role in fomenting public hostility is often attributed to environmentalist groups and movements, and also to economic interests

M. Bucchi, *Beyond Technocracy: Science, Politics and Citizens*,
DOI 10.1007/978-0-387-89522-2_1, © Springer Science+Business Media, LLC 2009

which exploit the resistance to specific innovations: for instance, the growers of organic product in the case of GM foods. "Scientific illiteracy" and "anti-scientific prejudice" are two of the epithets most frequently used by technocrats to underscore the gravity of the situation.[1]

In the words of a group of Italian scientists:

> A spectre haunts Italy, a spectre which provokes alarm by predictions of catastrophe which terrorize the population. It preaches that science and technology are inimical to humankind and nature, and it incites hostility against science by exploiting baseless fears which fog the reason. This spectre is called obscurantism. It assumes various guises, but the most dangerous of them, because of their reactionary and irrational nature, are environmental fundamentalism and opposition to technical-scientific progress.[2]

What is to be done, therefore? The remedies proposed by technocrats to deal with this supposed anti-scientific prejudice operate on two main levels.

First, given the knowledge deficit among citizens and politicians, complex issues should be decided by those who have the capabilities to do so: in short, the experts. This line of argument has been pursued to provide a couple of Italian examples, by the well-known oncologist and former Minister of Health, Umberto Veronesi, who has proposed "an upper chamber for ethics and science" or by the economist Giulio Sapelli, who suggests the creation of a "council of sages" (Veronesi 2003; Tiliacos 2004).[3] Akin to this perspective are the frequent calls for politicians to pay more attention to the opinions of experts and less to those of the man in the street.

Second, in order that experts may receive suitable social support in their task, it is suggested that medium- and long term initiatives should be devised to narrow the knowledge gap between experts and the general public. Such schemes should change public attitudes towards science and its activities, or at least attenuate the hostility against them. This emphasis on the public's inability to understand the achievements of science – according to a model of linear, pedagogical, and paternalistic communication – has associated the label of "deficit model" with this technocratic view of the public understanding of science.

Of course, this "missionary" branch of the technocratic position has ancient origins: consider, for instance, the activities of the Royal Institution in early nineteenth-century England. Explicit worries about the relationship between public opinion and science were expressed by policymakers in the 1950s. This was especially so in the USA after the World War II, when a concern to bring the general public closer to science involved the scientific community, scientific journalists, and government agencies. For example, the document *Science and the Nation* (1949),

---

[1]One authoritative example among many is provided by the Biotechnology and Biological Research Council (UK BBSRC 1996: 2). For Italy, *see*, for example, the recent study by Bellone (2005).

[2]From the manifesto of the movement *Galileo 2001 for Freedom and Dignity of Science*, http://www.cidis.it

[3]*See also* the Venice Charter drafted on the occasion of "The Future of Science conference", http://www.veniceconference2005.org and which speaks of "a permanent authority for Science".

issued by the Association of Scientific Workers, stressed the need to improve the public understanding of science by using conventional educational tools, the new opportunities offered by media like television, and the necessity that scientists, too, should "disseminate" scientific knowledge. The policy statement drawn up by the American Association for the Advancement of Science (AAAS) in 1951 – the so-called *Arden House Statement* – included among the association's goals that of increasing "the public understanding and appreciation of the importance and promise of the methods of science in human progress [..] in our modern society it is absolutely essential that science – the results of science, the spirit of science – be better understood by government officials, by businessmen, and indeed by all the people." (Weaver 1951, cit. in Lewenstein 1992: 52)

The joint schemes undertaken with the media for this purpose burgeoned to such an extent that in the early 1960s, the AAAS considered making "dedicated" offices in Hollywood and New York available to television producers (Lewenstein 1992). In 1958, six months after the launch of the Sputnik – an event which greatly encouraged investments in research and the promotion of scientific culture so that the USA could keep pace with its Soviet rival – the National Science Foundation (NSF) introduced a "Public Understanding of Science" programme costing one and a half million dollars. Finally, the National Science Writers' Association (NSWA) obtained funding from the Rockefeller Foundation to conduct a series of surveys on newspaper readers assessing whether they thought greater space should be devoted to scientific news. In 1957, a survey of the American public, again carried out by the NSWA, found attitudes that were decidedly in favour of science but modest levels of understanding; a result which was used to justify greater investments in science education programmes. It was for the purpose of appraising the impact of these programmes that, in the 1970s, the NSF included "scientific literacy" among its "science indicators".

However, it was not until the 1980s that an outright "movement for the public understanding of science" came into being. In Europe, a significant landmark was the report on *The Public Understanding of Science* published by the Royal Society in 1985 (Gregory and Miller 1998; Bucchi 2003a). According to this report, "better public understanding of science can be a major element in promoting national prosperity, in raising the quality of public and private decision making and in enriching the life of the individual." (Royal Society 1985, cit. in Irwin 1995:14). The report called for greater commitment in this regard by public and private institutions, specifying the benefits for the individual and society. As to the individual, people better informed about scientific subjects are able to make more informed choices in their daily lives (e.g., as regards health), besides being more fully able to appreciate the results of scientific research. As to society, if citizens are better informed about science and technology they will work more efficiently. By becoming less hostile and more amenable to research and innovation, they may facilitate political decisions and economic development and thus contribute more fully to the democratic process. The report concluded that "Scientists must learn to communicate to the public… and… consider it their duty to do so." (cit. in Gregory and Miller 1998:6). The document led to the creation by the Royal Society, jointly with the

Royal Institution and the British Association for the Advancement of Science, of a Committee for the Public Understanding of Science (COPUS) responsible, among other things, for allocating funds to schemes communicating science to the public. This interest in the relationship between science and public opinion led to a large-scale study on the public in Great Britain which achieved high visibility among scientists and policymakers (Durant et al. 1989). The survey found significant interest among the British public in science, an interest indeed greater than in sport and politics. However, a decidedly small proportion of respondents considered themselves sufficiently well-informed about science, and an even smaller, not to say negligible, number exhibited a minimally adequate level of factual knowledge about science. On numerous occasions, the results of these surveys have been used to argue that too little attention is paid to scientific topics and that the level of scientific understanding is poor. One of surveys most frequently cited – a comparative study between the USA and Great Britain – concluded, for example, that more than 90% of the American and English populations could be considered scientifically illiterate (Durant et al. 1991).

In light of the results these studies, various public and private institutions have launched initiatives and programmes to promote greater awareness and better understanding of science among the non-specialist public. These schemes range among the open days now organized by the majority of research laboratories and institutes, Festivals of Science, the initiatives mounted by foundations and companies expressly to "combat scientific illiteracy", and training courses for science journalists.[4] At the European level, besides the "European Science Weeks" instituted in 1993, mention should be made of the specific funding available under the European Commission's Fifth Framework Programme (1998–2002) "to raise the public awareness of science". Since the Sixth Framework Programme, the promotion of dialogue between science and society has been included among the European Commission's priority objectives for research. Various national and international initiatives have also been undertaken to encourage scientists to communicate with the public, and to provide them with the skills that they require to do so. Documents such as the Wolfendale Committee's report to the British Office of Science and British Technology (1995) have recommended that all publicly financed researchers should devote part of their time to illustrating their results to the public. For some years, in countries like Sweden, institutional obligations of the staff of public universities have included, besides teaching and research, the so called 'third mission': communication and engagement with the general public.

According to the recent Recommendation by the Committee on the European Charter for Researchers and Code of Conduct for the Recruitment of Researchers:

---

[4]For an international survey of these initiatives, *see* OECD (1997) and European Commission (2002).

> Researchers should ensure that their research activities are made known to society at large in such a way that they can be understood by non-specialists, thereby improving the public's understanding of science.[5]

1.2 The Flimsy Pillars of the Technocratic View

Both the components of the technocratic response, strengthening the scientific role of expertise and informing the public better so that it is more receptive to the instances of science, have significant weaknesses.

I shall begin by examining the major shortcomings in the "missionary" part of this response. I shall then argue that it is not just because of these shortcomings that the technocratic option does not seem feasible as a solution to the deadlocks which frequently arise in technoscience.

Essentially, the technocrats argue that the public's open scepticism concerning questions such as the introduction of transgenic foods is due to misinformation. As political scientist Renato Mannheimer has claimed, the almost general opposition to GM foods among Italian citizens recorded by the Ispo/AC Nielsen survey, as well as by Eurobarometer surveys and numerous other studies, is due to "media popularization and the image which the mass media have conveyed, rightly or wrongly, of GM foods. This has persuaded the majority of Italians to be opposed to them" (Mannheimer 2003).

If this is how matters stand, the remedy appears simple: it will be sufficient to communicate the results of scientific research more and better. Thus, the public will be better informed and will *consequently* embrace the positions of the scientific community. In turn, a more favourable "climate" of public opinion will strengthen the role of scientific research and this, via its technological outcomes, will produce economic growth and thus benefit society as a whole.

Hence, the "patient" (public opinion, the media) is given – often with the support of large public funding – huge injections of visits to museums and science centres, science festivals, open days at research laboratories and institutes, training courses for science journalists, television programmes, and quiz shows for prizes (!).[6]

Yet it has never been proved that the more the science is communicated, the more the level of public awareness increases, and even less so that attitudes towards specific scientific and technological issues grow more positive as a consequence. In general, moreover, although the problem of the effects of communication has engaged the attention of mass media researchers for more than half a century, the direct influence of the information media on attitudes (not to mention behaviours) is still largely to be proven.[7]

---

[5] http://www.europa.eu.int/eracareers/pdf/C(2005)576%20IT.pdf

[6] A list of the projects financed under the Fifth Framework by the European Commission can be consulted at ftp://ftp.cordis.lu/pub/improving/docs/rpa_projects_fp5.pdf

[7] *See*, for example, DeFleur and Ball-Rokeach (1989); on the inadequacy of a transfer model with specific regard to the communication of science, *see* Bucchi (2004a).

In the case of biotechnologies, various studies have shown that misinformation on its own cannot be considered as responsible for public hostility to biotechnological applications. Not only are people most exposed to sound scientific information by the media still critical of biotechnologies, so too are those best informed about them (Bucchi and Neresini 2002; *see also* Gaskell and Bauer 2001; Bauer and Gaskell 2002).

For that matter, the stubborn refusal to quit by millions of smokers shows that it is not enough to communicate a scientific fact, however indisputable it may be ("smoking seriously damages your health"), automatically to obtain significant changes in behaviour.

The second pillar of the technocratic discourse holds that public opinion, especially in Italy, is viscerally hostile to scientific research, little interested in acquiring information on scientific topics and mistrustful of researchers. But to dispel this stereotype, it is enough to look at one of the recurrent surveys conducted at European level on the relationship between European citizens and science. This reports a generally high level of trust in the scientific and medical professions and their practitioners: 71% of Europeans believe that doctors form the most trustworthy occupational category, followed by scientists (45%), and well above the other occupations. Italians, indeed, express a level of trust slightly higher than the European average: in particular, they are among the most optimistic that science and technology will be able to resolve problems like poverty, famine, and damage to the environment, and in general, they believe that the benefits of science and technology outweigh their potential risks (Fig. 1.1) (Eurobarometer 2001, 2005).

In regard to biotechnologies, recent years have seen universities and research institutes achieve significantly greater credibility among the public – Italian even more than European – to the point that it today largely outstrips that of environmentalist organizations and consumer associations (Observa 2005; Bucchi and Neresini 2006).

But it is above all in a different respect that the case of biotechnologies epitomizes the weakness of the stereotype of a public adamantly hostile to science. In Europe, but even more strikingly in Italy, surveys report strong public resistance against agro-food biotechnologies, but at the same time, a marked openness towards medical biotechnologies: over 90% of Italians, for example, believe that research should continue in the latter sector (Observa 2005).

It is evident that if the public were to reject biotechnologies because of prejudice against science, it would also reject biotechnologies in both agro-food and medicine. And even in the specific case of GM foods, the opposition of public opinion seems difficult to explain in terms of prejudice and "ideological" antagonism. In fact, whilst it is true that the majority of citizens are sceptical about current applications of agro-food biotechnologies, it should be remembered that almost 60% are in favour of research continuing in that area as well.

The third pillar of the technocratic account is misinformation. But is the public really so ignorant about science and technology? In absolute terms, it is undeniable that public opinion, not solely in Italy, exhibits significant gaps in its scientific knowledge. By now, proverbial is the finding that around one-third of Europeans

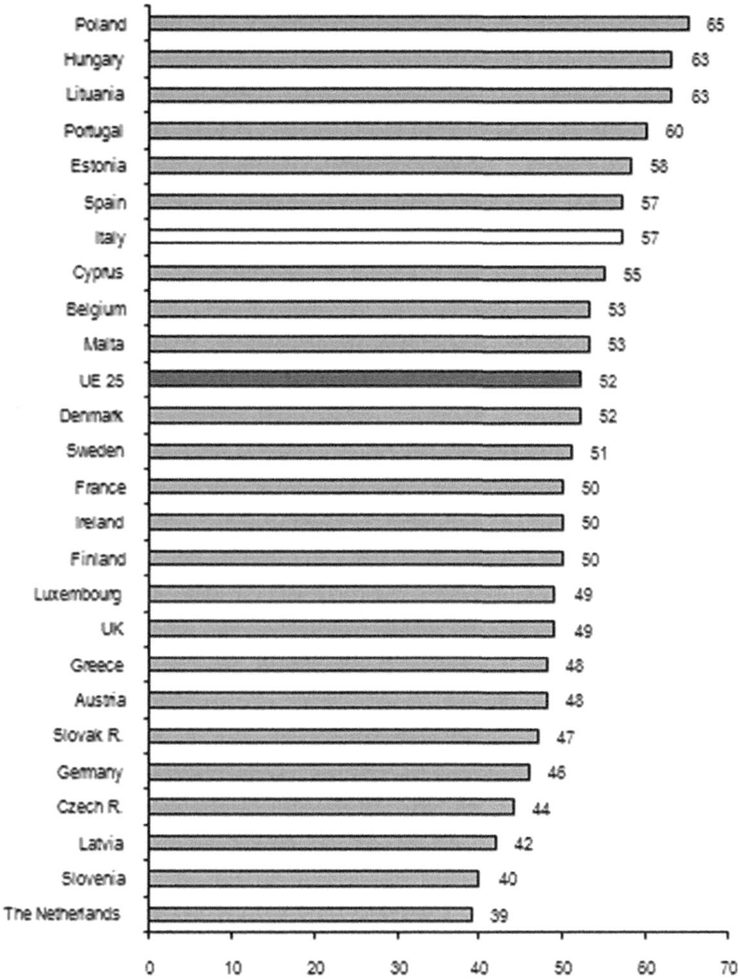

Fig. 1.1 Citizen's optimism regard science and technology: Benefits of science are greater than any harmful effects it may have (European Commission 2005:58)

believe that ordinary tomatoes do not contain genes whereas only transgenic tomatoes do so (Eurobarometer 2003; Observa 2005).

Moreover, what is meant by "ignorance" is by no means clear. Indeed, the indicators used to measure the public understanding of science have often been considered debatable. A 1991 study by the NSF reported that only 6% of the interviewees were able to give a correct scientific answer to a question about the causes of acid rain, but neglected the fact that experts themselves still disagreed on such causes. Another finding often cited to assert the public's "scientific illiteracy" is the large number of people who believe that astrology is a science. But the most recent data show that in this case, too, the problem is mainly terminological. If "astrology" is replaced with the expression "casting horoscopes", the number of respondents who consider it to be a scientific activity greatly diminishes, from 41% to 13% in the most recent European survey (Eurobarometer 2005:36).

Other studies have shown the complex array of images of science held by the public. A perception of astrology as a scientific discipline – classified by numerous surveys as an indicator of scientific illiteracy – is not infrequently accompanied by a good understanding of science (Wynne 1995). Such studies also take it for granted that terms like "theory" and "experiment" – indeed "science" itself – evoke the same unambiguous meanings for both the general public and the scientists.

More generally, it has been pointed out that the equation between the public understanding and the ability to answer questions about science has long restricted discussion of the public understanding of science to the somewhat tautological statement that members of the public do not reason in the same way as professional scientists do. This prompts the question of whether many surveys on scientific literacy do not instead measure "the degree of the public's social conformity to a stereotype held by scientists of a 'scientifically literate public'" (Layton et al. 1986, cit. in Wynne 1995:378).

A group of electricians working at the Sellafield nuclear reprocessing plant in the UK gave the researchers various reasons for their lack of interest – contrary to what one would expect – in acquiring scientific information about the risks of radiation. First, the electricians believed that interesting themselves in the scientific aspects of radiation would have caught them in a chain of pointless argument and discussion. Second, they feared that being confronted by uncertainties and probabilistic estimates of risks would cause them alarm, or even panic, and would therefore be dangerous. Third, the electricians said that there were other workers at the plant who possessed the information; any active effort on their part to acquire it would have undermined the trust and authority relations established in the workplace (Michael 1992).

In other cases, scientific information may be ignored by citizens as irrelevant to their needs, or simply because they distrust the source, believing it to represent interests other than their own. Thus, "technical ignorance becomes a function of social intelligence, indeed of an understanding of science in the sense of its institutional dimensions" (Wynne 1995:380; emphasis in the original).

The differing perceptions of experts and public cannot be explained solely by the disparate amounts of information available to them. It may instead be due to a more complex disjunction between expert and non-expert knowledge. A classic example of the gap between expert and lay knowledge is provided by the "radioactive sheep" crisis which erupted in certain areas of Britain in 1986, following the Chernobyl nuclear plant accident in Russia. British government experts long minimized the risk that sheep flocks in Cumberland had been contaminated by radiation. However, their assessments proved to be wrong and had to be drastically revised, with the result that the slaughter and sale of sheep was banned in the area for 2 years. The farmers for their part had been worried from the outset because they had direct knowledge based on everyday experience (which the scientific experts dispatched to the area by the government obviously did not possess) of the terrain, of water runoff, and of how the ground could have absorbed the radioactivity and transferred it to plant roots. This clash between the abstract and formalized estimates of the experts and the perception of risk by the farmers caused a loss of confidence by the latter in the government experts and their conviction that official assessments were vitiated by the government's desire to "hush up" the affair (Wynne 1989).

According to some scholars, experts themselves reinforce the representation of the public as "ignorant". During a study on communication between doctors and patients in a large Canadian hospital, a questionnaire was administered to patients in order to assess their level of medical knowledge. At the same time, the doctors were asked to estimate the same knowledge for each patient. The three main results obtained were decidedly surprising. Whilst the patients proved to be reasonably well-informed (providing an average of three correct answers for every four questions), less than half the doctors were able to estimate their patients' knowledge accurately. Finally, this estimate was in any case not utilized by the doctors to adjust their communication style to the information level that they attributed to the patients. In other words, the fact that a doctor realized that a patient found it difficult to understand medical questions or terms did not induce him/her to modify his/her explanatory manner to any significant extent. The patients' lack of knowledge – the authors of the study somewhat drastically conclude – appeared in many cases to be a self-fulfilling prophecy, for it was the doctors who, by considering the patients to be ignorant and making no attempt to make themselves understood, rendered them effectively ignorant (Seagall and Roberts 1980).

The fourth pillar of the technocratic thesis claims that if the public is misinformed about science and technology, it is the fault of the mass media. Mention has already been made of the difficulty of establishing solid correlations, and not only in regard to science, among exposure to information, attitudes, and behaviour. Two decades of studies on the media coverage of science prompt a series of considerations. First, it is simply not true that the media devote little space to scientific topics. An analysis of the coverage of science by the Italian daily press across 50 years has shown, for example, that the column inches allotted to scientific topics have significantly increased over time. A very similar tendency has been reported by long-period analyses of the daily press in the UK, Germany, and Australia (Kepplinger 1989; Australian Science indicators 1991; Bauer and Petkova 2005). Nor does it appear that the media depict scientific issues in a primarily negative or alarmist light: the majority of articles in the daily press treat science in positive terms, although it is undeniable that attention has recently shifted to the possible implications – in terms of risks to society – of certain developments in research and technological innovation (Bucchi and Mazzolini 2003; Bauer and Petkova 2005).

It should also be borne in mind that researchers and scientific institutions themselves often perform a leading role in such coverage. Around one-fifth of the articles on science published by the *Corriere della Sera* in the period 1946–1997 were written by scientific experts, whilst in Great Britain, one-quarter of the articles published in the daily press were based on press releases issued by research institutes. This phenomenon of pressure applied by "public relations" on science information is growing apace, and it will be discussed more thoroughly in the next chapter (Einsiedel 1992; Hansen 1994; Bauer and Gregory 2007; Bucchi and Mazzolini 2003; Goepfert 2005).

Finally as regards scant public interest in science, suffice it to point out that 70% of Italians believe that the media should give more coverage to biotechnologies – although these have certainly not been neglected in recent years – and they would like to know more about biotechnologies, not only through the traditional media but from public meetings with researchers, or by receiving information leaflets from national and local public institutions. In the UK, 79% of citizens, with a sharp increase in recent years, want scientists to spend more time discussing the implications of their research with the public (Observa 2005; Mori-Ost 2005; Bucchi and Neresini 2006).

1.3 Democracy and Ignorance

Even if the above-discussed pillars were more robust than have I tried to show, is it so important for citizens to be well-informed about issues concerning science and technology?

I have said that, although in absolute terms the public is probably rather ignorant about technoscience, it is important to consider the phenomenon in relative terms. And not just because, as said, Italians are at least as ignorant (or aware) as Europeans or Americans. not to mention the Japanese!

Considering the matter in relative terms also means relating the awkward questions raised by technoscience – GM foods, stem cells, and disposal of radioactive waste – to the other issues now firmly on the public agenda. Who could deny that a better-informed and more interested public would be desirable? Not just on science and technology, however. On how many important issues on the public agenda, can we form an opinion and decide on its basis whether, for example, to join a protest demonstration, to vote yes or no in a referendum, or simply to choose a candidate in elections? About how many of them do we have accurate and detailed knowledge?[8] Some time ago, on the occasion of an important intergovernmental conference in Rome attended by the European heads of government, a survey found that the majority of Italians knew very little about the themes of the conference, related to the European Constitution, even though these had been widely reported by the newspapers for days and had received close attention from commentators.[9] Yet no one would dare argue that ordinary citizens, as non-expert and misinformed, should be kept distant from decisions concerning the European Union or change to the electoral system, or reform of the Workers' Statute – questions which citizens

[8]On the relationship between democracy and the Utopia of an "omnicompetent" citizen, *see also* Lippmann (1925).

[9]*Internazionale*, 17 October 2003:26.

are often called upon to adjudicate in referenda. Moreover, nobody has ever considering allocating even a small part of the resources spent on communicating science to furnishing more detailed and clearer information about the contents and the reasons behind the European Constitution or the complex workings of the Italian electoral system.

Yet numerous comments – on the 1987 referendum to decide whether Italy should abandon nuclear energy, or the recent one on assisted reproduction – stress the inability of ordinary people to understand such complex issues, almost as if to suggest that on similar occasions, at the entrances to the polling stations, checks should be made not only on voters' electoral documents but also on their knowledge, perhaps by means of a special questionnaire or interview.

In its deficit model version, the mission of the technocrats seems bound to fail because it is impracticable, largely groundless, and in certain respects anti-democratic.

Let us imagine that it is possible to remedy the notorious "deficit" and bring citizens to a level of awareness judged acceptable. In this case, we would be immediately confronted by serious problem: who establishes what constitutes an "acceptable" level? Will it be necessary to create a special committee of experts to do so? And what is to be done if there is no consensus within the scientific community, as often happens with controversial issues? Besides this point, to which we shall return below because it is crucial, let us also imagine that it is possible to equip citizens with an "acceptable" level of awareness about GM foods and thus resolve conflicts on the issue. The very next day, it will be necessary to start again by informing the public about nanotechnologies, for instance, and the day after about nuclear energy, and so on.

Those who argue that we should pursue the Utopia of an "expert" public forget the reasons why they first conceived that Utopia: the increasing specialization of scientific research and therefore its perhaps tragic, but inevitable, counterpart: the abysmal ignorance which characterizes not only the man in the street but also an experienced researcher in high-energy physics when asked about research on embryo stem cells.

It is not surprising that we are ignorant about most of the issues on the public agenda. But it is indeed surprising that such ignorance – intrinsic to decision-making processes in contemporary societies and especially visible when such processes concern technoscience – has not to date prompted serious discussion in terms of epistemological status. Of course, by "ignorance" is meant not just the technical-scientific ignorance of the non-expert but also the ignorance of experts about the social, economic, and cultural implications when, for example, an innovation leaves the laboratory and interacts in sometimes unpredictable ways with other innovations (Hacking 1986; *see also* Chap. 2, Sect. 8).

No less Utopian than converting citizens into scientific experts (on all subjects, moreover!) is the frequently expressed ambition to alter science coverage by the media, where for the technocrats "alter" often means "get us a better press".

How is it believed that media coverage can be altered? Essentially by acting upon the journalists who write on science. This accounts for the numerous courses, laboratory visits, and training opportunities offered by institutes and associations of researchers – often with the contribution of the national or international public institutions – to journalists so that they can improve their knowledge about science and technology. Also in this case, however, the result appears anything but a foregone conclusion. First, because these opportunities are taken up by specialist journalists or individuals who deal routinely with science: the editors of science supplements in newspapers, popular science magazines, or television science programmes. But we well know that journalists of this kind are already amenable to the positions and arguments of the scientific community, to the point that they see themselves as attentive "interpreters" for public opinion of the scientific community's successes. And we also know that their messages reach a minority audience more interested and more "socialized" to the scientific community's points of view, when it has not already been substantially "trained" by that community (Peters 1994; Bucchi 2002). The risk, in short, is that the exercise will be pointless. A very different matter is reaching those journalists who concern themselves with science only when issues of general public interest arise – mad cow disease, GM foods, radioactive waste, and stem cells. During the 1997 BSE emergency that shook Europe from Great Britain, most of the reports in the daily press and on television news bulletins were not filed by science journalists, but by correspondents in London. Besides the banal consideration that such reports reach a broader section of the public, to be noted is that such journalists, unlike their "scientific" colleagues, feel a great deal less "loyal" to the world of research. If anything, they regard it as their duty to express the disquiet of public opinion, in whose service they do not hesitate to question or criticise the work of scientists. Not by chance, one of demands most frequently accompanying the dissatisfaction with media science coverage expressed by researchers and commentators is that such coverage should be "restored" to science journalists or brought under their closer control. But this demand appears not only contrary to the declared intent ("technocrat" researchers and commentators voice concerns about the possible "ghettoization" of science topics in the media); it is also impracticable because it runs counter to one of the main trends in the current scenario of science in society by exchanging the cause with the effect. The increasing "migration" of scientific issues from specialist columns and features to the news pages, and their corresponding transformation into public issues, is not a cause of the proliferation of conflicts on technoscience but rather one of its most evident outcomes.[10]

Moreover, the endeavour to convert the man in the street into a science expert seems largely unjustified. The urgent need to expand the flow of public communication of technoscience is generally taken for granted and rarely argued for convincingly. The chain "more communication = more understanding = more social support for science = more innovation = more economic development" is by now repeated like a mantra. Yet not the least doubt is expressed concerning the resilience of a series

[10]For an empirical analysis of this migration – or as a researcher has called it the "de-ghettoization" – of technoscience issues from dedicated to "generalist" contexts in the Italian and German daily press, *see*, for example, Bader (1990) and Bucchi and Mazzolini (2003). Analyses of television coverage are rarer: for a longitudinal analysis (concerned mainly with risks) of Italian television news bulletins, *see* Bucchi (1997).

of causal connections, each of which is questionable to say the least and where the linkages appears extremely fragile. But why should it be particularly important for citizens to be as informed about technoscientific matters as they are about the other issues on the public agenda? As said, the original aims of the movement for the public understanding of science developed at three levels. At the economic level, the intention was to create a climate favourable to research, and hence innovation, and thereby foster the well-being of society. Analysing this assumption would be beyond the scope of this book, however. I merely point out that the relationships among research, technological innovation, and development are by no means taken for granted by researchers. Several studies have disputed both that innovation necessarily and exclusively ensues from research (*see*, e.g., Rosenberg 1982; Faulkner 1994) and that it automatically translates into greater economic competitiveness (*see*, e.g., Comin 2004). It is even more doubtful that the prerequisite for all this is public interest and competence in science. Consider the case of Japan, where some of the world's highest levels of investment in research (3% of national wealth) and of innovation are matched by indicators of public science interest and information below the averages recorded in the USA and Europe (Nistep 2004).

Frequently invoked on the political level is the "civic duty" of citizens to inform themselves more thoroughly about technoscientific matters, so that they can participate in decision making. Postponing detailed discussion of decision-making processes to a later chapter, to be stressed here is the significant distance between the right to have certain information – should the citizen deem it opportune – and the duty to be informed at all costs. It is in the direction of the former right that significant national and international has been directed in recent decades – consider law 241/1990 on the transparency of administrative documents and access to them[11] – and it closely concerns the technoscientific sector. This is the case of European regulations on biotechnologies, or of the so-called "Seveso Directive", which establishes the right of residents in areas at risk of technological accidents to access information about potential dangers, or as amended in 1988 to place clearer emphasis on the right to know – more than the need to know – of the population concerned.[12] And it is the former right of information, and not the burden of receiving pre-packaged messages, that, according to the most recent surveys, is also deemed crucial by citizens so that they can deal with contemporary technoscientific dilemmas (Observa 2005; *see also* Chap. 4).

There is finally the cultural justification of enabling citizens to enjoy the richness and beauty of science. Here the objection is not so much against the intent as the method. For if this is the assumption, it is difficult to understand why the efforts and economic investments devoted in recent years to communicating scientific themes via the mass media has not been matched by a comparable concern, either nationally

[11] Recently amended by law 15/2005. On this, *see*, for example, Arena (2001).

[12] This concerns respectively Directive 2001/18/EC of the European Parliament and Council, of 12 March 2001, on the deliberate release into the environment of genetically modified organisms, and Directive 501/82 and relative amendments of 10 November 1988. On the latter case, *see also* Dini Valentini (1992).

or internationally, to foster basic science education, and particularly the teaching of science in schools. Yet, if there is a factor which unquestionably influences the ability to understand the issues raised by technoscience, it is basic education (especially scientific) (*see* Bucchi and Neresini 2002, 2006); just as poor schooling (and not anti-scientism) appears to be a concrete weakness affecting Italy in international terms. Moreover, it is at this level that it seems reasonable to take concrete action on problems such as the "decline in scientific vocation" – the steep fall in enrolments on science degree courses (especially mathematics, physics and chemistry) – apparent not only in Italy but also in Europe, the USA, and Japan for some years (*see* Observa 2004). Not only in this respect may the rhetoric of the struggle against anti-scientism and scientific illiteracy redound against more urgent needs regarding research policies, removing resources from specific problems and using them instead for a generic "battle of civilization". Furthermore, this insistence on communicating and publicizing science risks lumping together very different, and sometimes contradictory, priorities: the dissemination of research results, developing an interest in science among young people, achieving greater visibility for institutions and specific researchers, or strengthening the role of scientific experts in decision making. Emblematic in this regard is the prominence given in numerous countries to science's role in investigative work through explicit communication initiatives or, indirectly, through successful television series like *Crime Scenes Investigation*. Whilst this visibility has certainly generated enthusiasm among young people, increasing enrolments on university courses in forensic science, it has also created problems for experts in the sector. In fact, exposure to science's decisive and unproblematic role in the investigations which is typical of television series like *Crime Scenes Investigation* may, for instance, perplex the members of a jury when scientific evidence is presented (as often happens in these cases) in cautious and probabilistic terms (Hooper 2005a).

Conversely, situations which the technocratic-paternalistic doctrine would unhesitatingly call "public misunderstanding" may give rise to greater visibility for science and assist specific areas of research. Consider, for instance, the misconception based on so-called "genetic determinism", which holds that there is a specific gene responsible for biological or even behavioural traits, from laziness to homosexuality. This is a misconception which experts deprecate as a gross distortion of increasingly sophisticated genomic knowledge, but it nevertheless maintains an extraordinary hold on the public. Yet it is difficult to believe that decisions like that of the US Congress to allocate huge resources to the project of mapping the human genome, removing them from large infrastructures for high-energy physics, or in general the high visibility enjoyed by that project, have not been bred by a determinist conception largely predominant among non-experts.[13] According to some scholars, a striking

[13] The biologist and Nobel Prize winner Walter Gilbert would introduce his lectures on the sequencing of human genes by showing a compact disk to the audience and saying "This is you": *see* Nelkin and Lindee (1995), Fox Keller (2000), and Lewontin (2000). On the case of genetics as an example of the inadequacy of a simplistic model of communication based on "transfer", *see* Bucchi (2004). For one example among the many of how "genetic determinism" has been stigmatized by experts, *see* Jordan (2000).

example of a "misrepresentation" with great efficacy in the public arena is Einstein's theory of relativity. The appropriation of this concept by twentieth-century culture, as in the commonplace saying that "everything is relative", was frequently in open contradiction with the intentions of Einstein himself: whose aim, in fact, was not to relativize every point of view but, on the contrary, to demonstrate that all the important laws of physics can be written in such a way that they have the same form for any inertial observer and therefore an "absolute" value independent of that observer (Sparzani 2003).[14] This was in certain respects a colossal "misunderstanding" but it probably fit well with a series of expectations held by the culture of the time, and it helped make Einstein into a twentieth-century icon. Emblematic is the photograph of him sticking his tongue out, which was also celebrated by the semiologist Roland Barthes as epitomizing the image of genius and of physics as the science *par excellence* in terms of public perception.

Finally, there have also been frequent clashes between the "missionary" discourse and certain features of public debate in democratic societies. I have already mentioned the tendency for transparency (accessibility of information when necessary) to be confused with communication. In the context of the mass media, for instance, it is not rare for journalists, especially "generalists", to perceive offers to improve their knowledge of scientific matters as encroachments on their freedom and professional independence. During the mid-1990s, the Ciba Foundation, financed by the pharmaceutical company Ciba (today Novartis), created a free telephone consultancy service for science journalists called the *Media Resource Service*. By calling a telephone number, journalists could state the topic on which they wanted to write and be referred to an experienced scientist who would give them information. The introduction of the service aroused numerous protests among journalists, who considered it an attempt to manipulate one of their essential prerogatives – the selection of sources – which responded to criteria entirely different from those of the scientific community. Reactions of the same kind have sometimes been provoked among the general public when institutes and practitioners in the scientific and technological sector, public institutions, or economic actors have launched schemes to improve public knowledge about topical technoscientific issues. Some participants have believed that the real purpose of these initiatives is to persuade citizens to look more favourably on the research or technological applications in question (Pur due 1999; Irwin 2001).

Of course, this is not to imply that every attempt to communicate science is useless or bound to fail. The examples cited are instead intended to illustrate critical aspects

[14] Moreover, as well known, Einstein's 1905 article published in *Annalen der Physik* with the title "On the Electrodynamics of Moving Bodies" did not contain the expression "theory of relativity" which subsequently made him famous throughout the world, even among non-experts. The expression was introduced by Planck and subsequently adopted by Einstein "even in awareness of its inadequacy" (Sparzani 2003:246, n.6). In his 1916 article on general relativity, Einstein refers to "that theory designated today as the 'theory of relativity'" (cit. in Sparzani 2003:247). On the role of the public debate in the discussion of Einstein's theory, *see* Biezunski (1985).

of a view of science communication typical of the technocratic discourse, especially in its paternalistic "deficit" version. The first of these critical aspects is the belief that information is not an end in itself or a civic right, but a means to induce specific attitudes and behaviour in the public. The second is the assumption that knowing more about the scientific issues on the public agenda necessarily means accepting the value-judgments of experts and therefore enthusiastically supporting all their proposals or innovations. One of the risks is that the "scatter-shot" approach advocated by this view may lead to the neglect of more solid policies designed for the long period and of more demonstrable efficacy, for example, better-quality science education which uses a wide range of teaching tools, from multimedia resources to interactive science centres.

But if both the bases of the "missionary" vision of the diffusion of scientific culture and its effectiveness are debatable, how is it possible to explain its persistence, also among authoritative commentators and exponents of the scientific community, to the point that it assumes the features of an outright "ideology"? The question is all the more important if one considers that a similar "stimulus-response" communicative model ("I communicate to you, hence you are convinced") would provoke outrage in any sector (politics or religion) other than science.

Recently, when commenting on the unsatisfactory results of an exhibition on embryo stem-cell research organized by the Barcelona Science and Technology Park (on leaving the exhibition, visitors seemed not to have changed their opinions on such research), one of the curators concluded: "Unfortunately, it is very difficult to get people to change their minds." (Malagrida et al. 2004). *Fortunately*, it is very difficult to get people to change their minds! Otherwise the public debate would be too easily manipulated by those in a position to administer large doses of communication.

Why, therefore, do so many of the actors concerned stubbornly adhere to a view of the relationships between experts and public opinion which is weak in both its premises and results? Even the most convinced supporters of the technocratic standpoint cannot have failed to notice that after 20 years of "pedagogical" initiatives for the public understanding of traditional science, conflict on technoscientific has increased, and not decreased.

A first reason is the special status attributed to scientific discourse. According to one view (in truth more widespread among the public than among experts[15]), the scientific standpoint – allegedly "objective" and concerned with the "truth" unlike, for example, a political platform – can impose itself merely by being communicated, and immediately overcome any resistance provided it is understood. The weakness of this belief does not result from the intrinsic goodness – historically a matter of much epistemological discussion – but to its decline in terms of public

[15]As numerous studies have shown, the conceptions of "science" present among researchers are often anything but epistemologically naive (Chia 1998). But this does not gainsay that the scientific community has selectively assumed this public image – according to the well-known process by which public communication removes caution and contingent aspects from scientific statements so that they become irrefutable and intuitive certainties (Fleck 1935; Whitley 1985; Bucchi 1998a).

perception. Rightly or wrongly, the scientific discourse has gradually lost its privileged status, being largely assimilated to the other viewpoints which feed into the public debate.

Another reason concerns the specific historical-professional context in which this "missionary" conception has come about. Because preoccupation with the inability of the public to deal with scientific questions has developed within the scientific community, it is easy to understand the temptation of scientists, certainly not typical of them alone, to identify problems and solutions outside their remit. From this point of view, apparently ingenuous is the claim of some sociologists that the public understanding of science movement should problematize scientific activity at least as much as it problematizes the other two terms of the question – the media and the public. Precisely because of the difficulty of questioning oneself and one's role in society, scientists have been induced, not necessarily in a conscious way, to conceive diagnoses and solutions which shift responsibility for changes onto society. The technocratic recipe for resolving conflicts between science and society is thus elementary: change society – make it more "educated", more rational, more "disciplined" – to use the term in Foucault's sense (Foucault 1975), in order to render it more compatible with science. This also helps clarify the apparent paradox between the technocratic response on the decisional level – "all power to the experts" – and on the pedagogical one – we work for a more "informed" public. Resolving the paradox is the expectation that public opinion, merely because it pays more attention to experts and is better informed by them, will endorse their positions. Citizens must indeed be given information, but only as much as they need to understand that they must bow to the experts.

In Chap. 2, I shall seek to show that this endeavour has failed not only by default but also by excess. It has been shown, in fact, that just as it is implausible to inject large doses of information to align public opinion with the beliefs – for that matter rarely unanimous – of experts, so it is implausible to steer the public's involvement once it has been activated. It has happened that some groups of citizens have informed themselves much more than expected by technocratic paternalism, and with results at odds with the latter's expectations.

It must of course be acknowledged that the scientific community's capacity to dictate the agenda of relationships between science and the public has been due – besides the authoritativeness that it has enjoyed in the public arena – to its ability (indubitably heightened by the feeling that its position is being threatened) to recognize, well before other actors, that something is changing in those relationships.

It should also be stressed, and as mentioned at the outset, that adherence to a technocratic point of view does not entirely coincide with technoscience and with the views of technoscientific experts. For reasons discussed in more detail later, the recipients of technoscientific expertise, such as policymakers and citizens, have often actively contributed to sustaining the technocratic option with their expectations, or because they have been attracted by opportunities to delegate responsibilities.

Finally, it would be ungenerous not to acknowledge that quite a few experts themselves have warned about the flaws in the technocratic thesis. At least at international level, there is less and less talk about the need to combat scientific illiteracy and increasing emphasis on encouraging, not the diffusion of scientific ideas, but active public engagement and dialogue between science and society. This should replace the simple transfer of ideas and attitudes from one side to the other. In 2001, for example, the European Commission changed the name of its programme for financing projects in this sector from "Public Awareness of Science" to "Science and Society". In 2000, a report by the British House of Lords announced that the phase of traditional public understanding of science, based on one-way, top-down communication, had come to an end. The report called for new forms of involvement which were responsive to a new "mood for dialogue" in the public (House of Lords 2000). In 2002, the three institutions (Royal Society, Royal Institution, and British Association for the Advancement of Science) which in 1985 had jointly founded the COPUS decided not to renew the appointment of its executive, thus decreeing the end of its activities.[16]

Nevertheless, one suspects that still lurking behind such declarations of principle is the technocratic vision and its vocation "to convert the unfaithful", with all the strength and seductiveness of a model where it is only necessary to press a button (communication) to obtain the result desired. For example, initiatives which declare that they are founded on dialogue and debate at the same time declare that their aim is to persuade the public to relinquish its resistance against biotechnologies or nanotechnologies.[17]

Emblematic in this regard is the disappointment expressed by numerous scientists and policymakers with the results of the large-scale "GM Nation initiative", a broad debate on transgenic foods launched by the British government in 2003. When the significant scepticism of public opinion became apparent in the debate, the initiative was closed down, its failure being blamed on the domineering behaviour of organized groups, most notably environmentalists. I shall not repeat here my earlier comments on communicative expectations: dialogues or debates are by their nature liable to generate conflict, or at least differences of opinion. But it is worth noting that it is inevitable in public debate that the most organized participants, even if they are not the most representative, gain greater visibility and influence. The same happens when trade unions representing only a minority of workers sit at the negotiating table or reach agreements on behalf of all the workers in a particular sector.

An idea of the widespread persistence of the conventional view of society as resistant to advances in science is conveyed by the recent conclusions of an influential working group of European scientists:

> The one lesson to emerge after a decade of controversies (GM food, stem cells, reproductive technologies) is that research, development and innovation can hardly prosper in the face of social opposition to science (European Group of the Life Sciences 2004).[18]

---

[16] "We have reached the conclusion that the top-down approach which COPUS currently exemplifies is no longer appropriate to the wider agenda that the science communication community is now addressing": joint statement by the three institutions, 6 December 2002, http://www.copus.org.uk

[17] As reported, for example, by the Director of *Ecsite*, the European network of Museum and Science Centres: *see* Staveloz (2002).

[18] http://europa.eu.int/comm/research/life-sciences/egls/pdf/conclusions_egls.pdf

1.4 A Flour that Threatened to Bring Down a Government

On 4 August 2000, the then Italian Prime Minister, Giuliano Amato, announced a decision by his cabinet which had repercussions both in Italy and abroad. Almost one year earlier, in September 1999, an environmental association, *Verdi Ambiente e Società*, had lodged a lawsuit against the marketing authorization issued for seven starches and flours derived from genetically modified maize and rapeseed by the Novartis, Monsanto, Pioneer, and AgrEvo companies. The lawsuit disputed the authorization of these products obtained by means of the so-called "simplified procedure". According to this procedure, allowed by the European regulations in the case of foodstuffs "substantially equivalent" to conventional ones, a company can obtain permission from the competent authority of one of the European countries to market a product throughout the European Union. The seven starches and flours had in fact been authorized by the Advisory Committee on Novel Foods and Processes of the British Ministry of Agriculture. The environmental association contested the applicability of this procedure to the products in question, disputing their "substantial equivalence", the absence of which would require the more laborious ordinary procedure which involves tests and authorizations by individual national authorities.

The matter had already been considered by the Minister of Health in the previous government, Rosy Bindi, who had asked the *Istituto Superiore di Sanità* for an advisory opinion. The institute's conclusion was that the products were substantially non-equivalent to conventional ones. In December 1999, another opinion was issued, by the *Consiglio Superiore di Sanità*, which reiterated that the starches and flours in question could not be considered "substantially equivalent", and therefore advised that application of the simplified authorization procedure was illegitimate. In the meantime, after the resignation of the government headed by Massimo D'Alema, Umberto Veronesi had been appointed as Minister of Health, and the Green Party's Alfonso Pecoraro Scanio as Minister of Agriculture. At the beginning of July 2000, Veronesi asked for a second advisory opinion from the *Istituto Superiore di Sanità*, which confirmed its previous findings. The *Istituto Superiore di Sanità* responded to a further request from the Minister by stating that it had given a restrictive interpretation to the concept (considered unclear) of "substantial equivalence" and concluded that it was "not obliged to express an opinion concerning the risk of possible environmental release of the GMOs in question and on their products". During a stormy cabinet meeting, the Minister of Agriculture Pecoraro Scanio and the Minister of the Environment Ronchi threatened to resign if marketing of all the products was not immediately suspended. Of contrary opinion were the Ministers of Industry and of Research, whilst the Minister of Health refrained from joining the discussion.

On 4 August 2000, the President of the Council announced the suspension of four of the seven products. Since this was the first time that the so-called "safety clause" had been applied for products of this kind – some other GMO varieties approved by the ordinary procedure had instead already been suspended by member-states – it was necessary to consult the European Scientific Committee on

Food, which would decide whether the conditions were in place to suspend marketing of the products throughout Europe. The European Committee decided that the documentation submitted by the Italian authority did not demonstrate that there were significant risks to human health.

In the following months, hostility against GMOs in Europe grew apace both in a considerable number of member-states and the European Parliament. In Italy, the decision of the Minister of Agriculture to make access to ministerial funding conditional upon an undertaking by researchers to discontinue experiments with GMOs provoked protest among the scientific community. More than a thousand researchers, among them the Nobel Prize-winner Renato Dulbecco, signed a petition "for the freedom of research", which was published in the daily newspaper *Il Sole 24 Ore* on 5 November 2000.[19]

Complex political decisions, contradictory opinions by authoritative research institutes and committees of experts, economic interests, protests by environmental groups: the story of the transgenic flours at the time of the Amato Government has all the ingredients for it to be considered emblematic of contemporary dilemmas in technoscience. Emblematic but not unique, however; for a few years later, in 2003, the head of the Piedmont regional government, Enzo Ghigo – but Lombardy and Veneto followed closely behind – found another awkward dossier on GMOs on his desk. The health authorities had discovered that around 300 ha of unauthorized GM crops were being grown in the region, and it was feared that they might contaminate conventional crops: a question much discussed by experts. After heated debate among researchers, companies operating in the biotechnology sector, and representatives of the farmers' associations, Ghigo opted for the "zero tolerance" approach and ordered the crops in question to be destroyed.

The technocrats make an elementary diagnosis of cases of this kind: besides obtuse and hostile public opinion, the problem is due to a political class "deaf" to the reasons of science. As the philosopher of science Riccardo Viale puts it, the error of politics in such cases is that it "refuses to consider institutional science – as expressed by the principal international science journals – as the sole source of knowledge about physical and biological phenomena. It therefore implicitly accepts that the choice of the sources of knowledge and its modes of production are driven by social and cultural factors" (Viale 2003). On this view, the political class, seeking to gratify public opinion for electoral purposes, pays insufficient attention to the counsels of science. The argument is interesting because it reproaches politics for not working like science but instead following its own institutional rules, which are more attentive to social and cultural reasons that to those of peer review.

The technocratic position seems to address, and claims to resolve, the contemporary dilemmas of technoscience by invoking a return to a traditional pattern of relationships among science, politics, and society. But it forgets that these dimensions and their relationships have changed. And they have not changed only in the sense that political problems with high technoscientific complexity have increased.

---

[19]For a detailed reconstruction of the affair, *see* Meldolesi (2001).

This is only the surface of the problem. Also the politicians of the past were confronted by issues of this kind: consider the decision by the government of the USA to construct and use the first atomic bomb. The difference is that the relationship between politics and scientific expertise, as well described by C. P. Snow, used to come about "behind closed doors", sheltered from public scrutiny (Snow 1960). Why for many years was there no public discussion, as conversely happens today in the case of GMOs, of whether it was dangerous to use herbicides or fungicides in agriculture? Were not these innovations just as potentially controversial?

Today, however, this "closed" arrangement, which partly resembles the one advocated among others by Viale and Veronesi (i.e., a politics which resorts to scientists as advisors and meticulously follows their instructions), is no longer practicable, for various reasons.

The first of them is the role of the media that makes it impossible to conduct the relationship behind the closed doors of power. Against the hackneyed view that the media are mere tools in service of the powerful, it is easily argued that whilst the spread of a mediatized public sphere – and all the more so, the diffusion of the new media – has created visibility opportunities for the leading players on the public stage, it has also made control of such visibility unprecedentedly difficult (*see*, e.g., Meyrowitz 1985, especially chapter ix; Thompson 1998). This difficulty does not concern only the personal images of political leaders[20] and public actors (including scientists), who are subjected to close scrutiny with their inadequacies being pitilessly laid bare. It also affects decision making itself and the role played in it by technical-scientific expertise. Cases like that of Andrew Kelly, the English scientist who committed suicide in 2003 after media investigations had revealed his controversial involvement in compilation for the Blair government of the "dodgy dossier" on Iraq's weapons of mass destruction, highlight the extent to which decision-making centres are now penetrable.

The second reason is the transformation of scientific expertise, and above all its perception by public opinion and politics. Today, the question is not so much whether to trust the scientific experts as which experts to trust. Politics, even more than public opinion, is assailed by a host of experts who frequently offer conflicting advice. The environmental organizations have their own trusted scientific experts, who act as their mouthpieces when the greenhouse effect or GMOs are discussed. What, therefore, should the "responsible politician" invoked by authoritative commentators do when faced with a decision on, say, destroying or preserving transgenic crops? Conduct a straw poll among researchers dealing with such matters? Consult recent issues of *Nature* which has published studies denying the dangerousness of GMOs but others which do not rule it out? Believe, as in the case of the Amato Government and the transgenic flours, the first opinions of the *Istituto Superiore di Sanità* and the *Consiglio Superiore di Sanità* or the second opinions

[20] A classic example: George W. Bush surprised in September 2005 by a close-up television image showing him writing a note to ask Secretary of State Condoleeza Rice if a session of the United Nations could be suspended so that he could go to the lavatory.

discoveries skyrocket the quotations of private companies (*see* Chap. 4; Gibbons et al. 1994; Etzkowitz and Webster 1995; Ziman 2000). Indeed, for some time, there has been lively debate among experts themselves on the problems raised, for instance, by the publication of results of research funded by pharmaceutical agro-food companies, and on resistance against disseminating such results when the research outcomes prove harmful (*see*, e.g., van Kolfschooten 2002).[23]

Also increasingly responsible for the eclipse of science's image as disinterested and immune to material interests is the frequent activism of researchers, with appeals and street demonstrations. Both these phenomena will be analysed in the next chapters.

[23] Subsequently, a large group of scientists signed a protest letter against the failure of the journal *Nature Biotechnology* to report a conflict of interest concerning the authors of an article that argued for the innocuousness of GM foods. The letter stated that 11 of the 18 authors had received or were receiving funds from companies operating in the agro-food biotechnologies sector.

Chapter 2
Einstein Has Left the Building: Coming to Terms with Post-academic Science

Mr. Halley presented his planisphere,
with a short description,
to his majesty, who was very pleased;
but received nothing but praise.

(John Aubrey, Brief Lives)

"Albert Michelson,[1] hearing of Einstein's confirmation of his findings, strode into the Royal Society in a tight black T-shirt printed with the legend 'Michelson Rocks'. Contrary to rumour, Michelson has not made a fortune from his shares in Interferometers-R-Us.com, though others have. He sold too late and made only 'a few million', he told me, just enough to buy his famous 90-foot yacht." (Collins 2005:50).

This, according to Harry Collins, is the account that we would have read in the newspapers at the time of Michelson and Einstein if the styles and organizational practices of technoscience had been the same as they are today. It is an ironic way to measure the distance that separates contemporary technoscience from its not too distant past. It is difficult to conceive how inapplicable the traditional technocratic interpretative frame is to the dilemmas and impasses of technoscience unless one takes account of the profound changes that have taken place in scientific research, and in its interaction with the broader social context. In necessarily schematic manner, this chapter analyses some of the main factors responsible for such changes.

2.1 A Post-academic Science?

Although their views do not always coincide, numerous scholars, as well as numerous leading researchers, maintain that the science of today differs profoundly (and not only in organizational terms) from the science that we used to know: namely,

[1]Albert Abraham Michelson (1852–1931), American physicist of German origin. Known above all for having obtained, in 1882, an extremely accurate experimental measurement of the speed of light, and for the experiment with which he showed, together with Morley, the absence of any effect of the speed of the earth on the ether, the substance whose existence had been postulated in order to explain the propagation of light waves. Michelson was the first American physicist to be awarded the Nobel Prize (1907).

DOI: 10.1007/978-0-387-89522-2_2,

academic science, and especially the "big science" which arose in the industrialized countries during the inter-war period. This was a science driven by large research institutions (mainly public) and large investments (also mainly public) and based on a solid fiduciary relation between political power and a narrow circle of experts.

This was the science that contributed to the success of the Allies in the World War II, and in regard to which heads of government like Churchill took crucial decisions upon consultation with their scientific advisers.

This was the science in whose name Einstein could write to President Roosevelt urging him to support the work of American nuclear physicists so that the USA could develop a nuclear bomb before Germany; and on conclusion of the war, to write again to Roosevelt recommending that he heed the warnings of Szilard, a nuclear physicist, about the risks of a war fought with the recently developed nuclear weapons.

This was the "goose with the golden eggs" which Vannevar Bush described in his 1945 report (*Science: The Endless Frontier*) to President Truman spelling out the equation: "more basic research = more technology = greater prosperity = greater ability to keep pace with the enemy in times of cold war".

It was above all physics – big science *par excellence* – that required broad international collaboration agreements in order to sustain the costs of its experiments on elementary particles using enormous accelerators.

This was the science addressed by the first real research policies and subsequently – especially from the end of the 1960s onwards – by the first public debates on the role of science in development and the environment. It is from this science that the configuration of contemporary technoscience appears to have distanced itself in numerous ways. Indeed, scholars now talk of a "post-academic science", or "mode-2 science", contrasting it with the "mode-1 science" that arose in the post-war period, and some commentators have even identified a "second scientific revolution" comparable to the one that generated modern science in the sixteenth and seventeenth centuries (Gibbons et al. 1994; Nowotny et al. 2001; Etzkowitz and Webster 1995; Ziman 2000).

2.2 After Doctor Strangelove: How I Learned Not to Worry and Love the Stock Exchange

A first difference is indubitably represented by the financial aspect. In numerous countries and technoscience sectors, particularly since the 1990s, private investments have supplemented government funding and in some cases compensated for cutbacks in the latter. Agreements on co-operation and patenting between universities and companies have become increasingly frequent, and they have sometimes been actively encouraged by policymakers. In the 1980s, for example, the federal government of the USA began to off-set reductions in funding with permission for universities and their researchers to patent the results of their research. The situation in Europe is more complex, even if collaboration between companies and research centres for years has been one of the European Commission's priorities in research policies implemented to revive European competitiveness as envisaged by the Lisbon Strategy.[2] Recent increases in investments have propelled countries like Finland and Sweden to the top of the European and worldwide table in terms of resources devoted to research. And these resources have derived in large measure from relations with companies (like mobile phone manufacturers) whose investments cover more than 70% of the total costs of research and development. Overall, more than half of the 677 billion dollars spent on research and development around the world derive from investments by multinational companies.[3]

There have been international funds investing in medical biotechnologies for a number of years. Numerous firms, so-called "spin-offs", have been set up jointly by universities or research institutes and private organizations to exploit research particularly promising in terms of industrial innovation. No less common, in certain sectors especially, is the presence of scientists on the boards, or among the main shareholders, of companies operating in the technoscience field. More and more often, the announcement of a scientific discovery boosts the stock of the company that has invested in that particular project, or commercial interests significantly impact on the processes of communicating and sharing traditionally central to research. After the sensational cloning of "Dolly, the sheep", in 1996, PPL Therapeutics – the company which had funded the work of Ian Wilmut and the Roslin Institute with around ten million pounds – imposed silence on all the researchers involved in the project until patent applications had been filed for a cloning technology to breed animals able to produce therapeutic proteins in their milk (Kolata 1997).

This increasingly close interweaving between business and research attracted public attention worldwide on 14 May 2000, when the English Prime Minister Tony Blair and the then president of the USA issued a joint declaration stating that data on the human genome, including the human DNA sequence and its variations, should be made freely available to scientists everywhere (Danchin 2000). The next day, the stocks of numerous biotechnological companies plunged, dragging the NASDAQ technological index down with them.

What had happened? It had happened that Celera Genomics, Inc., a private American company supported by the funding and technology of another company, Perkin-Ellmer, in its turn associated with the computer colossus Compaq, had announced that it was about to complete the mapping of the entire human genome. At the head of Celera Genomics was Craig Venter, a biologist who before setting up on his own with 70 million dollars of private funding had long worked on the huge project of sequencing the human genome at the National Institutes of Health. Thus, a recently created private company was on the brink of accomplishing what a vast international consortium of public research institutes had been striving to achieve for decades. Worries that such an important scientific result, and one so

---

[2] http://europa.eu.int/growthandjobs/pdf/lisbon_en.pdf

[3] Figures referring to 2002 from the United Nations Conference on Trade and Development: *see* UNCTAD (2005).

appetizing in terms of financial gain, might be removed from the public domain sparked heated controversy and induced the researchers of the two groups to agree on joint publication of the human genome "map". The map was indeed published in 2002, and in the two most prestigious international scientific journals – *Nature* and *Science* – but with a slightly paradoxical inversion whereby the article by the public consortium was published in the privately owned journal (*Nature*) and that by Celera in the journal of the American Association for the Advancement of Science (*Science*).

Especially areas such as microelectronics, information technology, and biotechnologies exhibit an intersection between research and the market as well as the "transformation of scientific knowledge into economic activity" (Etzkowitz and Webster 1995:482). Besides bringing new actors with their legitimate interests – entrepreneurs or company shareholders – onto the technoscience stage, this change has redefined the role of the scientist, replacing the traditional "division of labour" which rewarded the researcher in terms of reputation and the industry in terms of profit. Thus, corporate consultancy by university researchers, once deemed "external" to their role, may in this new setting be an integral part of their research and the tasks assigned to them by the institution, and indeed a factor enhancing them. On the contrary, the traditional processes of evaluating work by colleagues through peer review may conflict with the need for secrecy entailed by the possible commercial exploitation of research. And this augments the centrifugal pressure towards private consultancy and funding. In the medical and pharmaceutical sectors, the role of private financiers has become so pervasive that frequent disputes arise within and without the scientific community on "conflicts of interest", or resistances against publishing results contrary to what the commercial sponsor expected. In the USA especially, the impact of this change has profoundly altered the institutional and organizational setups of universities according to their ability to access market-sourced funding. Indeed, universities themselves have become influential economic actors. In budgetary terms, fully seven American universities appear on the *Fortune* list of the 500 companies with the largest gross revenues (Table 2.1). Columbia University on its own in 2003 had 52 of its own companies, 169 research agreements with other private companies, and 133 million dollars a year of revenue from its patents. This new role, moreover, has made university administrators increasingly the key actors in decision-making processes which used to be dominated by academic staff (Brint 2004; Washburn 2005).

Of course, this is not to demonize the role of companies or of commercial interests; for without them, numerous important discoveries and innovations, from the times of Liebig until the Genome Project, would probably not have been achieved. Without the interests of the French cattle farmers, Pasteur would not have been able to conduct some of his most important experiments on vaccines. And without economic interests we would not have the aspirin or the soup cube today. Moreover, it is well known, and especially to scholars concerned with the processes of technological innovation, that interests play a significant part in the process whereby some technically viable innovations rather than many others are "selected" (*see*, e.g., Bijker 1995).

Table 2.1 American universities in the *Fortune* list of the top 500 companies, 2003

	Position	Annual budget (billions of dollars)
University of California	113	18.1
Harvard University	273	6.9
Stanford University	350	5.0
Yale University	396	4.2
MIT	419	4.0
Duke University	459	3.6
University of Michigan	491	3.3

Source: Institutional Date Archive 2003, cit. in Brint 2004:3

The point is understanding how the change in the sources of funding reconfigures the organization of scientific practice, as well as redefining the professional identity of researchers and the perception of technoscience by the public. As the Physicist and Philosopher of Science John Ziman has written, "the contrast with academic science could scarcely be sharper" when we are confronted by a science which is "Proprietary, Local, Authoritarian, Commissioned, and Expert". This is a science that produces knowledge which is not necessarily made public, centred more on local technical problems than on general understanding, controlled by managerial authority, and commissioned in view of practical goals from experts required to solve concrete problems rather than demonstrate their creativity (Ziman 2000:78–79). Merely denying these changes, as is not infrequently done by leading members of the scientific community – "The majority of scientists do not even know what multinationals are" declared the Nobel Prize-winner Levi Montalcini on Italian television with reference to the mobilization of researchers against restrictions on GMO research – inevitably fuels the belief that the relationship with business is necessarily harmful to scientific activity.

2.3 Whose Knowledge?

The change in question, together with other factors, has driven and complicated the debate on the nature of intellectual property which has significantly, but not exclusively, traversed technoscience in recent years. One of the crucial issues in this discussion is, of course, the question of whether the results of scientific research can be patented. Contrary what is often believed, this discussion did not suddenly begin with the advent of sectors like biotechnologies. Instead, its roots extend deeply in the history of science and technology.

The concept of the "individual paternity" of a scientific discovery, which to us today seems obvious, was not in fact such in the second half of the nineteenth century, when there arose a "heroic ideology" of discovery and the invention as the fruits of genius and individual inspiration (*see* Boyle 1996, esp. chap. 6; MacLeod 1996).

Previously just as widespread had been a "deterministic" rhetoric holding that discoveries and inventions derive from the maturation of a certain strand of inquiry or from the impelling needs of an industrial sector; the role of the specific scientist or the inventor was merely accidental. This conviction was embraced in particular by intellectuals with more egalitarian ideas: in 1767, the chemist Joseph Priestley even went so far as to criticise the attribution of the term "genius" to scholars like Isaac Newton, claiming that this practice obscured the true nature of scientific inquiry as "patient and industrious" teamwork (cit. in MacLeod 1996:149).

The clash between the two ideologies came to a head during the following century, and especially in Great Britain, where it dragged on for many years amid lawsuits and polemics in the press between the "determinists" contrary to any type of patent and their adversaries instead striving to make protection easier to obtain and more restrictive. A number of factors were probably responsible for the victory of the heroic-individual vision of discovery and invention, which could therefore be brought under the protection of patents: the success of the first great international expositions showcasing the most recent technological advances, the advent of the biography as a literary genre, and the more general need of civil society, in numerous countries, to flank religious saints with secular figures of high symbolic value. It is certainly significant that those same years saw the spread of biographies and monuments dedicated to inventors and scientists. Particularly celebrated is the statue of Faraday in the lobby of the Royal Institution (1876), whilst outstanding examples in Italy are the monuments dedicated to Volta (Pavia 1878), Galvani (1879), and Giordano Bruno (1889).[4] Some years later (1900), the Nobel Prize was instituted in accordance with the will of the inventor and entrepreneur Alfred Nobel; a prize which would become, for public opinion as well, the epitome of creativity and individual genius in science. But the deterministic conception of discovery as the fruit of tradition and teamwork did not disappear, instead, it was adroitly alternated with the heroic version (also by scientists themselves) according to the specific communicative situation, so that a repertoire of "cultural maturation" was privileged in discourse among experts whilst a repertoire centred on "genius" was emphasized in communication to the great public (*see* Brannigan 1981; Bucchi 1998a).

However, during the late 1900s, the issue of intellectual property acquired significant ramifications. Mention has already been made of the central importance assumed by biomedical research. The rapid growth of this sector led to the application, too hastily, according to some commentators, of a patenting system originally intended for engineering innovations (and subsequently extended to chemicals) to biology, so that also living organisms could be patented (Tallacchini 2003). In the USA, after a first phase of uncertainty and the rejection of patent applications by the US Patent & Trade Office in the 1970s, a series of rulings by the Supreme Court

[4] *See* Eberle (1997). It has been pointed out that the iconography of scientists and inventors in this period was largely modelled on the religious iconographic tradition (Mazzolini personal communication).

culminated in the controversial decision on the "Chakrabarty case" (1980) which definitively opened the way for patents in the biotechnology sector. The decisions of the Supreme Court established two principles which had fundamental implications. First: the distinction between inorganic and organic substances had no legal basis; hence, a micro-organism could be patented just as much as a new type of bottle cork. Second: the decisive difference was not between engineering and chemistry or biology, but between what existed and what did not exist in nature. Consequently, Einstein could not have patented his formula $E = mc^2$, whereas a biologist could patent a micro-organism that she or he had isolated and purified. Following these rulings, with the backing of a report by the Office for Technology Assessment, it became the practice of the American courts to remove the onus of proof from discoverers/inventors: these did not have to demonstrate that whatever they wanted to patent did not already exist in nature; rather, it was the office or court contrary to the application which had to prove that it did exist. The tacit equivalence established between complex and simple living organisms led in 1988 to the patenting, by two geneticists at the Harvard Medical School, of the OncoMouse™, a transgenic mouse carrying a human tumour gene which could be used to study breast cancer. Registered, with a view to its commercial exploitation, in the names of Harvard University and the multinational Du Pont, which had made unlimited funds available to the two university researchers, the OncoMouse™ was the first patented animal in history.

The patenting of living organisms has provoked numerous controversies and judicial disputes in recent years. The marketing of patented transgenic seeds has been fiercely contested by some sectors of public opinion as a potential cause of the impoverishment of farmers and their dependence on the large multinationals producing those seeds (see, e.g., Shiva 2001). No less controversial is the issue of the patent protection of the results of large-scale surveys on a population's genetic heritage contracted out to private companies. An example is provided by Iceland, where the government was harshly criticized when the company in question, whose shares had been purchased by numerous citizens, entered severe financial crisis. The patented existence of the OncoMouse™ has had mixed fortunes outside the USA. The European patent was granted after a turbulent procedure and a legal challenge brought by a British animal rights association; it was instead rejected by the Supreme Court of Canada. In 1990, the Supreme Court of the USA adjudicated the case of John Moore, a patient whose spleen had been removed at a university clinic California. From Moore's spleen, the researchers had obtained and patented a cell line able to produce proteins with anti-cancer properties. The Court ruled that Moore was not entitled to a share of the profits deriving from the patent (Wilkie 1993).

The issue of the intellectual ownership of technoscience products has become particularly topical as a result of the incentives provided by the research policies of numerous countries. In the USA, approval of the Bayh-Dole Act has actively encouraged public-sector researchers to patent the results of their work. In 1990 and 1991 alone, one-third of all patents issued in the USA since 1969 were granted to American universities. Underpinning (though this is not always made explicit) these measures has been the conviction that an opportunity to profit materially from their work is a powerful incentive for researchers, which is a new and more dynamic version of the link among research, innovation, and development theorized by Vannevar Bush.

Specific mention has been made of opposition against the patenting of living organisms due to worries concerning the violation of life and nature, the risk of their enslavement to the machinery of capitalism and the market, and standardization of their rich diversity for the purposes of mass production, which also has repercussions in terms of cultural identity (Harraway 1997).

At a more general level, those who criticise (particularly from the inside) this drive towards the patenting and protection of technoscience products instead argue that it is deleterious to the development of knowledge. The fear of missing opportunities for profit may hamper the circulation of ideas which has been the distinctive and essential feature of the development of modern science.

Such worries also concern the accessibility of research results, an issue which has grown increasingly controversial in recent years. In this case, the roots of the problem can be traced to the seventeenth century, when scholars found that they were spending more and more time on keeping abreast of the work of others by purchasing books and exchanging correspondence. Henry Oldenburg decided that he would do the task for them on a commercial basis. He founded the journal *Philosophical Transactions of the Royal Society* (1665), and although not himself a scientist, he selected and summarized the papers sent to him by the foremost scientists of the time (*see* Price 1963; Bazerman 1988).

In fact, the dissemination by scientists of their work among fellow researchers has been regulated by what sociologists and anthropologists of science have called a "gift economy" (Hagstrom 1982). Researchers cede articles presenting their results to journals for free, and then equally freely distribute copies of them to their colleagues, for instance, during conferences. It is then necessary for someone, in exchange for a commercial return, to print and distribute those results. These are the publishers of the specialist journals which over time have become the fulcrum of the contemporary scientific debate, especially in the natural sciences. The number of these journals has grown apace. It has been estimated that there are currently around 20,000 of them in circulation, with over 2 million articles published every year. Their increasingly marked sectoriality, the concentration of publishing groups into what closely resembles an oligopoly, and other peculiarities of this market (especially in some areas of research) – low-price elasticity, because consultation of certain journals is essential for researchers to be able to do their work, and the fact that it is not researchers but the libraries of their institutions that pay – have led in recent decades to a sharp increase in the costs borne by research institutions in acquiring such journals. According to some estimates, whilst the cost of books in the USA increased 4-fold between 1970 and 1990, in line with the cost of living index, the prices of specialist scientific journals increased 12-fold in the same period (Butler 1999; Scanu 2004).

Furthermore, the protection of intellectual property by patents acquits scientific publications from performing a crucial function: not so much exchanging information with colleagues – an "incidental" aspect according to the historian De Solla Price[5] – as claiming priority for results (Price 1963:68).

It was in this scenario that in 1994, Steven Harnad, a psychologist at Southampton University, sent an e-mail to some of his fellow researchers setting out what he called a "subversive proposal". Why, asked Harnad, must we pay not once but three times to keep abreast of work by our colleagues? This material is first paid for by the universities and research institutions that disburse our salaries, it is then paid for by the funds allocated to research projects by public and private institutions, and then it is paid for a third time when we purchase – again through our institutions – the subscriptions necessary for us to read the journals which publish these results. Not to mention the fact that we do not receive payment either when we publish articles or when we lend our time and abilities as referees. The growth of the cost of the subscriptions not only risks bankrupting our universities, it may also hinder a process fundamental for all researchers: publicly demonstrating to colleagues their priority on the results obtained, since patents cannot perform this function in all sectors.

It was these considerations that prompted Harnad's subversive proposal: what he called "skywriting", or the use of the internet by researchers to make their results freely available to all colleagues who might be interested. The incoming bottleneck – the selection of articles – should be kept as narrow as possible, whilst the exit bottleneck – the number of readers and conditions of access – should be as wide as possible. The peer-review system with which the scientific community guarantees the quality of publications should be rigorous, but the weight represented by the profit seeking of commercial publishers should be removed. As a model to emulate, Harnad pointed to the electronic archives already existing in some disciplines, for example Arvix, a database created in 1991 by the Los Alamos Physicist Paul Ginsparg where researchers in high-energy physics could post articles already accepted by journals but not yet published (so-called "pre-prints"). In just one decade, Arvix had made around 2,10,000 articles available for free online. Similar electronic archives were created in the following years for the biomedical sciences (*PubMed Central*) and the cognitive sciences (*Cogprints*), whilst open access journals published entirely on the web spread among various other research sectors as well.

In 2001, a group of American biologists headed by the Nobel Prize-winner for medicine, Harold Varmus, launched the *Public Library of Science* (Plos) in order to create a single database of scientific articles accessible to experts and non-experts alike. The initiative was rapidly joined by more than 30,000 scientists around the world. Today Plos is also a journal (*Plos Biology*) which is entirely similar to other specialist titles in terms of article selection by peer review but different as regards its financing. Because it is any case the research institutions which pay – the basic idea behind *Plos Biology* holds – we might as well eliminate the cost of consultation

[5]Price recalls that scientists like Kepler or Hooke not infrequently encrypted their results, thereby establishing their priority but without disclosing too much information to their rivals (Price 1963:68).

by directly soliciting contributions from the research institutions whose work we publish. Publication fees may be waived – thanks to the financial support of partners such as the *Wellcome Trust* and George Soros's *Open Society Institute* – in the case of research institutions unable to pay the charges. In October 2003, a group of European research institutions signed in Berlin on the proposal by the *Max Planck Gesellschaft*, a document which they undertook to recognize and evaluate papers published in open-access online archives and to encourage their own researchers to use such archives (*see* Laser Group 2005).

Although the impact of this phenomenon is still little studied, it certainly extends beyond the mere reduction of costs for research institutions. The existence of archives like Arvix or *Plos* has profoundly altered the everyday practices of numerous researchers. Visits to the library by researchers to consult the most recent issues of journals in their sector, usually published at intervals of many months except in the case of inter-disciplinary journals like *Nature* and *Science*, are now flanked (or indeed replaced) by the daily consultation of electronic archives of pre-prints.

However, this expansion in available information and the accelerated speed of its dissemination may have effects which theorists like Merton would have called "dysfunctional" for science. The possibility of accessing in almost real time the ongoing work of colleagues, and especially of rival research groups, is liable in certain cases – when it is discovered, for instance, that another group is close to the same result that one is working towards – to discourage continuing research which might instead prove fruitful on two counts: complementarity between approaches and methods to the same problem and the duplication of research which has frequently proved beneficial to scientific activity. Also to be borne in mind is that the same research may yield, as has sometimes happened, results which are different from those expected but nonetheless valuable (Merton 1957, 1961, 1963; Mitroff 1974). Moreover, however stringent the peer-review filter on these archives may be, there is an undeniable difference between the physical constraints of space on a paper-based journal with a pre-defined publishing schedule and number of pages – numerous journals have built their reputations on the small quantity of articles published compared with those received – and the virtually unlimited capacity of the internet.

A well-known distinction drawn by the historian Derek De Solla aids understanding of the extent to which the debate on "free access" to scientific publications and initiatives like *Plos Biology* concerns one of the cornerstones of academic science. De Solla described the change from "little science" to "big science" as an exponential explosion in the number of researchers and publications. According to his calculations, the scientists alive today represent almost 90% of all those who have ever lived. Price defines a scientist pragmatically as someone "who has published at least one article in a scientific journal", and science as what is contained in specialist journals (Price 1965:55).

If articles printed in specialist journals used to be the elementary measurement units of a researcher's recognition and reward, the metric determining his or her career progress, what counts today – and what will count in the near future – as a "publication" in post-academic science? An article archived in an online journal or database?

A pre-print on the researcher's web site? A deposited patent? An informal posting in an electronic discussion forum? Of non-marginal interest in the debate on access to research results is collective mobilization by scientists – a phenomenon, previously unknown, which has occurred repeatedly in recent years and is examined in Chap. 3.

2.4 From Physics to Biology

In terms of investments, ability to influence research policies, and public visibility, research in the era of big science was dominated by disciplines like physics. The advent of post-academic science has coincided with the irresistible rise of the life sciences, which use physics as their organizational and strategic model of reference. Physics had made a "quantum leap" from the "2 blackboards and 82 glasses" in Oppenheimer's car (which in the 1940s was all that the greatest American physicists needed to meet and discuss their research) (*see* Gleick 1992:4) to great international projects and complex investments in infrastructures. Likewise, in the space of a few decades, biology has moved from the shed where Watson and Crick worked on the research which led to discovery of the structure of DNA to the huge investments and the broad research groups involved in the Genome Project. A similar change of scale in the life sciences has driven the processes of interaction with industry described earlier. "Basic research costs hundreds of thousands of euros," explains Pier Paolo Di Fiore, head of the *Istituto di Oncologia Molecolare* of the *Fondazione Italiana per la Ricerca sul Cancro*, one of the founders, together with scientists and entrepreneurs, of the Italian biotech company Genextra; "the development of applications by biotechnological companies costs tens of millions of euros, and clinical tests by pharmaceutical companies hundreds of millions. The more you go up the scale of magnitude, the less space there is for non-profit organizations." (Dell'Oste 2005: 9).

This trade-off between the two sectors is often by marked a significant process of "transmigration" whereby scientists trained in physics move to biology. Corresponding to this transmigration is, according to some historians, the contemporary one of conceptual models and paradigmatic metaphors: in this sense, the gene is for contemporary biology – the equivalent of what the atom used to be for modern physics (Keller 1994).

At least until the mid-1970s, research in the life sciences seemed able to restore the "honeymoon" between science and public opinion ruined by the open conflict provoked by the use of nuclear energy for war and civil purposes. The end of the affair was depicted by Stanley Kubrick in his film *Dr. Strangelove, or how I Learned not to Worry and Love the Bomb* (1963), in which the stereotype of the scientist was already very different from the abstract Einsteinian genius. It had been replaced by the demented scientific advisor – modelled, according to some, on the physicist Edward Teller, one of the protagonists of the Manhattan Project to construct the first hydrogen bomb – whom not even political power was able to restrain.

The illusion of a new science yielding solely benefits had been cultivated until 1974, when a group of biologists signed a letter asking for research on recombinant DNA to be halted until its potential risks had been clarified. The primacy achieved by the life sciences in terms of public visibility is also very apparent from the long-period changes in their media coverage, according to what has been called the "medicalization of science journalism". In the Italian daily press since the 1980s, one article in every two on a scientific topic has dealt with biomedicine, and an identical tendency has been observed in Great Britain. The same applies to perception by the general public: in Italy, for example, it appears that "science" and "medicine" largely overlap for the man in the street (Bauer 1998; Bucchi and Mazzolini 2003; Borgna 2001).

Nevertheless, there are significant differences between the life sciences and the physics of the period of big science: for instance, the possibility to conduct significant research with limited financial resources and equipment. Whilst high-energy physics was entirely reliant on large investments, today, in sectors like the biotechnologies, large-scale projects co-exist with ones of more limited scope, but not for this reason devoid of implications, especially in practical terms. In 2003, great controversy was provoked in the international scientific community by the joint decision of some of the most prestigious international publications (among them *Nature*, *Science*, and the *Proceedings of the National Academy of Sciences*) to "alter or reject" scientific articles that might give potential bioterrorists information with which to produce deadly bacteriological weapons using simple equipment (Vos 2003).

In some situations, the availability of biomedical technology extends beyond the specialist domain. At the end of the 1980s, in the USA, and especially within groups of patients engaged in gay activism, it became common practice to distribute kits to perform self-diagnoses of HIV positivity (Epstein 1996). Today, at the cost of 280 euros, a credit card can be used to purchase online a DNA test with which to determine a child's paternity. The purchaser receives an absorbent kit on which to deposit a sample of saliva, returns it by post, and within 15 days receives the result (Chemin 2005).[6] By 2010, on the basis of a method developed at Harvard University, it may be possible for an individual to obtain by e-mail, at the cost of around 1,000 dollars, a DNA analysis warning of a possible genetic predisposition to pathologies like cancer (Nelkin 1994:29).

2.5 A Mediatized Science

Does anyone remember the case of cold fusion? What has become of the two scientists, Pons and Fleischmann, who featured on the front pages of newspapers around the world with their claim that they had obtained a nuclear fusion reaction by means of an unprecedented electrolytic procedure, a reaction that seemingly held out the promise of an extraordinarily cheap and clean source of energy? Even

[6]The web site where the do-it-yourself kit is on sale is http://www.dnasolutions.co.uk

before they had described their experiment, Pons and Fleischmann were harshly criticized by many of their colleagues for having announced to the press results not yet officially published in a specialist journal. Some 15 years later, news of research on cold fusion is increasing sporadic, but the use of the media as an amplifier, and sometimes even a springboard, for research and results has become routine.

Academic science used to despise the media on the grounds that they perverted scientific ideas among the general public, and that they were "dirty mirrors" reflecting an opaque and distorted image of research. It dismissed communication to the non-experts with the pejorative epithet "popularization" (*see* Friedman et al. 1986; Burnham 1987; Gunter et al. 1999). And after all, why should science bother to talk to the public when its relationship with politicians in the corridors of power was so close?

By contrast, post-academic science increasingly views the media as crucial interlocutors. Whether because of mistrust in the ability of communication to remedy the deficits in the public understanding of science, or because of osmosis between organizational models due the increasing interactions, as already discussed, with the business world, or because they have realized that good media visibility is something that political decision makers and financial investors increasingly want, the fact remains that all universities or research institutes now have public relations offices which organize press conferences to publicize those organizations' most significant activities. Not to mention the courses or guidelines provided to researchers so that they can deal with the media or at least avoid awkward situations: "If you feel trapped, obfuscate; it will get cut if it's too technical" is one of the recommendations on giving interviews that the *New England Journal of Medicine* makes to researchers (cit. in Nelkin 1994:29).

Aside from the missionary vocation of technocracy, these communicative efforts serve more pragmatic purposes such as gaining greater visibility and prestige for one's research. A former press officer at the *Royal Society*, one of the most prestigious associations of scientists, describes this change of attitude towards the media, thus "[the Royal Society] aims to use the media coverage to achieve the outcomes sought by the organisation" (Ward 2007: 159). It was on this basis that in 2005, the Royal Society launched a campaign addressed to journalists "to tackle misrepresentation of the scientific evidence on climate change in the UK national print, broadcast, and online media" (Ward 2007: 160). The campaign consisted of press releases, letters, and documents which rebutted point by point the arguments of commentators who minimized the seriousness of climate change, or it used a strategy of "preventive strike" whereby journalists were furnished with information intended to undermine the credibility of climate change deniers and pre-empt their media visibility. Thus, the *Royal Society* provided *The Guardian* newspaper with information that led to the publication of a front-page article revealing that the *Scientific Alliance Forum*, which was about to issue a report critical of climate change, was funded by a number of oil multinationals.[7]

[7] "Oil Firms Fund Campaign to Deny Climate Change", *The Guardian*, 27 January 2005, 1. *See also* Ward (2007).

In general, contacts with the media are no longer grudgingly conceded. On the contrary, research institutions actively pursue media attention and visibility. Already in the mid-1990s, one-quarter of the articles on scientific topics published by the English daily newspapers were based on press releases from research institutions. In recent years, the cutbacks in newspaper staff or the outright closure of numerous science desks have been matched by an increase in journalists and personnel employed in the public relations offices of research institutes and companies. It is estimated that there are currently around 60,000 journalists and around 20,000 employees working in public relations in Germany. In the USA, public relations officers (around 1,62,000) by now largely outnumber journalists (around 1,22,000). Newspaper journalism costs have been further reduced by the "information kits", ready to be turned into news stories, distributed by public relations officers. Indeed, it is estimated that around two-thirds of agency reports on scientific topics are based on press releases and other materials furnished by press and public relations officers. According to a broad study conducted on eight German national newspapers, fully 80% of articles on scientific topics are based on a single source, but fewer than one-third of them expressly mention the fact. Indeed, there are several examples of newspaper science pages being "outsourced" to the press office of a local university, as in the case of the German daily paper *Badische Zeitung* of Freiburg. In order to counterbalance the diminished media interest in the newsworthiness of science conferences, especially in the medical field, it is increasingly common for journalists to be invited, at the expense of the organizers and sponsors, so that they are induced to write about them. According to numerous researchers and science journalists, these developments (and particularly the scant transparency concerning the origin and use of public relations materials) are liable to undermine independence and critical capacity (Goepfert 2007).

Nor is fiction exempt from this "colonization" by scientific sources and institutions. For instance, the American Film Institute has recently organized, in collaboration with the Pentagon, seminars with scientists and screenwriters on producing films to persuade young Americans to choose science faculties when they go to university.[8]

The other side of this phenomenon is the fact that media exposure of techno-science no longer takes place after the scientific debate has settled a certain issue. Instead, media exposure comes about during the stage of greatest uncertainty and controversy among experts. The traditional linear sequence "research → informal discussion among colleagues → official publication of results → communication to policymakers → absorption into and stabilization by the disciplinary corpus through handbooks → popularization among the general public", which characterized scientific communication throughout the period of big science, is constantly broken up and reassembled.

These changes have been amplified by the speed of the new electronic communications media. The web typically disrupts the above sequence and diminishes the resilience of a series of "filters" which in the past controlled the progress of scientific results from the researcher to the general public, via a series of specialist forums.

[8] *The New York Times*, 4 August 2005.

A Google search on "applications of nanotechnologies" returns, already on the first page, specialist articles, advertisements, policy documents, enthusiastic opinions on the future of nanotechnologies, and worries about their implications (*see* Trench 2008). By joining discussion groups or mailing lists, any web user can follow controversies among experts once carefully concealed from non-experts, or read the positions of "orthodox" and "sceptical" scientists on a particular issue (GMOs, for instance).

The above-described pressures for open access to specialist publications have made available to non-experts – patients and companies – materials once accessible to them only in the libraries of specific institutions. In 2003, the *Royal Society* set up a working group which was tasked, among other things, with answering the question "what, if any, quality checks or filters should researchers subject their results to before communicating them to the public?" (Trench 2008:190). A similar question and a similar formulation would have been unthinkable a few decades ago. The capacity of peer review to act as a communicative filter becomes questionable when the highly competitive context of post-academic science combines with media allowing the rapid propagation of multidirectional communicative content outside the traditional forums.

Media exposure now impacts on every stage of the communication process. It short-circuits public discussion, specialist debate, and policy decisions, and in some cases, it even penetrates laboratories. It should be stressed that it is not solely due to a generic trivialization of journalistic information but also results from the increasingly strong pressure applied by the research community on the communications media. When, on the occasion of the Mars mission, which had been given enormous prominence in the international media, the director of the European Space Agency was asked why it was lagging behind NASA, he cited the role of their respective offices as a decisive factor. Besides being better-resourced and more experienced, the NASA press office could count on more rapid communication with the outside, whereas the ESA press office had to overcome numerous bureaucratic filters and controls.[9]

A general consequence is that the productive routines and the specific interests of the media have become increasingly influential in setting the technoscience agenda. Numerous studies have shown that researchers are attentive to media coverage of scientific topics: a paper published in the prestigious *New England Journal of Medicine* is three times more likely to be cited in the specialist literature if it has been mentioned in the *New York Times* (Phillips 1991). In 1998, the same *New York Times* gave great prominence on its front page, followed by large part of the information media around the world, to the discovery, by the Boston researcher Judah Folkman, of a revolutionary method to fight cancer by stopping the flow of blood to cancerous cells. The head of the American National Cancer Institute immediately declared that clinically testing the method was his institute's absolute priority, and the Nobel Prize-winner James Watson even proclaimed that Folkman "is going to cure cancer in two years",[10] comparing him with Darwin for his discovery.

---

[9]Cordis news, 17 February 2004, http://www.cordis.lu

[10]"Cancer: Drugs Raise Hopes for a Cure", *Herald Tribune*, 4 May 1998. *See also* Revuelta (1998).

The share value of Entremed, Inc., the company holding the patents for two proteins used by Folkman in his tests – angiostatin and endostatin – rapidly rose on the NASDAQ. Some time later, the author of the front-page article announcing the discovery, the science journalist Gina Kolata, said that she had already received a large advance from a publisher to write Folkman's biography. In the following years, however, no developments or significant confirmations of Folkman's results were forthcoming.

One sometimes finds, indeed, that scientific communication adapts to the rhythms and needs of media coverage. The project to map the human genome, for example, with its repeated announcements of partial, promised, or even imminent breakthroughs, ideally matched the need of the media for specific events to report, and without which a project lasting several years has scant newsworthiness. The result was fully 1,069 articles between 1996 and 2001 in the *New York Times* alone, with a peak of coverage which coincided not with the most significant scientific event – final publication in the specialist journals – but with the above-mentioned statement by Blair and Clinton in 2000, and the consequent promise that the ultimate goal would be achieved.

Dynamics typical of the scientific community and dynamics proper to the contemporary media also overlap in amplifying the personalization of technoscience. And this is not only in the sense that protagonists of technoscience – from Steven Hawking to Craig Venter – have become media stars. At the highest levels (Nobel Prizes, experts with particular prominence in the public arena), visibility becomes a credential that can be deployed not only when policy decisions are being taken but also in gaining privileged access to communicative arenas. Moreover, the more media resonance becomes an important resource for research institutions as well, the more it becomes probable that the public visibility of certain scientists heavily conditions the choices made in researcher recruitment, or in the definition of research priorities.

There thus arises a further paradox in the technocratic approach to communication: years of communicative efforts oriented by the deficit model have made science more sensitive to the needs of the media, rather than making the media sensitive to the needs of science.

2.6 A Science Without Boundaries

Whilst academic big science became institutionalized and developed thanks to increasing sectoral specialization, a distinctive feature of post-academic science is its tendency to break down boundaries, and primarily the traditional ones among basic research, applied research, and their technological implementation.

In key sectors of contemporary research like biotechnologies, nanotechnologies, information and telecommunications sciences, post-academic science is increasingly characterized by proximity between research and the contexts of its application (*see* Faulkner 1994; Gibbons et al. 1994; Ziman 2000; Nowotny et al. 2001).

This proximity is by now significantly recognized on the regulatory level – for example, American patent law makes no distinction between "invention" and "discovery" – as well as on the public perception that scientific research and technological innovation overlap. This explains why terms like "technoscience" are used to denote processes which, although indubitably distinct analytically, in practice are lumped together from various points of view – those of legal regulation, policymaking, the market, public opinion, as well as frequently by the protagonists of research and innovation themselves.

Also the boundaries among disciplines and specializations are being redrawn. Mention has already been made of the fertile interaction, in regard to concepts and human resources, with physics that has distinguished the growth of modern genetics. No less influential in defining a paradigm centred on the gene has been the concept of "information" developed within cybernetics (Keller 1995). Today, projects involving researchers and methods from different disciplines have become the rule rather than the exception, with the birth of declaredly inter-disciplinary sectors such as bioinformatics. Most of the problems addressed by contemporary research require the competences of numerous disciplines (in the case of BSE, e.g., veterinary pathology, neurophysiology, the microbiology, animal dietetics, epidemiology, and agricultural economics). Inter-disciplinary collaboration is increasingly encouraged by the institutions which allocate research funds not only within the natural sciences but also between the natural and social sciences, and especially in areas like BSE or biotechnologies. Inter-disciplinarity is now one of the strategies most frequently adopted and openly advocated to achieve excellence by the leading American universities, so that it flanks, and in certain respects supplants, the traditional strategy based on the recruitment of the best researchers (Ziman 2000; Brint 2004).

This permeability is often apparent in the boundary between scientific experts and non-experts, with technoscientific processes embedded in heterogeneous social networks comprising researchers, politicians, business people, journalists, patients, and even computer hackers. The peer-to-peer technologies developed for content sharing without a central server, like the Napster music-file sharing program, provided a model for the human genome researchers to coordinate the activities of the project's various sequencing centres (Merriden 2001). This increasing participation by non-experts – and citizens in particular – in technoscientific processes will be discussed more thoroughly later (*see* Chap. 3).

There is then the further type of permeable boundary, less metaphorical and more concrete than the previous ones discussed, which consists in geographical borders. It is undeniable that since its institutionalization, scientific research has been distinguished, well before and much more than other social activities, by its endeavour to transcend national barriers (*see* Rossi 1997; Merton 1942). Sociologists of science have coined the expression "invisible colleges" to denote the formation of communities made up of researchers physically distant but constantly in contact and actively engaged in the exchange of information, methods, and results (Crane 1972). The institution and subsequent administration of the Nobel Prize in neutral Sweden has always emphasized that science is "universal" and that it lies above political cleavages and alignments even in periods of fiercest conflicts among states – the two world wars or the cold war.

However, post-academic science construes this endeavour in more distinctive terms. On the one hand, the development of communication technologies has profoundly changed the practice of research, further weakening the spatial constraints on collaboration among research groups and favouring the sub-division of complex activities or the continuous monitoring of long-term experiments. The laboratory that used to embody academic science from the architectural point of view, to the point that the construction of a laboratory historically signalled the institutionalization of an intellectual territory and its disciplinary independence (Home 1993), has in many sectors partially dematerialized into networks and connections that do not necessarily require the physical co-presence of researchers in the same place. This loosening of spatial ties is reflected in dynamics that replicate, on a smaller scale, the broader socioeconomic processes of globalization. We consequently witness the shift of significant research and development activities to areas in which labour costs are lower. Thus, the constant connection guaranteed by the internet has enabled Portal Player, the Silicon Valley company which produces the microchips for the Apple iPod digital music player, to transfer large part of its research, development, and planning activities to Hyderabad in India.

It no less frequently happens that research activities are moved to areas where regulation of certain types of research is less stringent: this is the case, for example, of experiments in human cloning conducted in countries like Dubai and Korea. According to the 2005 annual report of the United Nations Conference on Trade and Development, research and development activities have become among the largest recipients of investment in the emerging countries, above all by multinationals. Between 1993 and 2002, expenditure on research conducted at the foreign subsidiaries of multinationals increased from 30 billion to 67 billion dollars. In the same period, investments by large companies in China, India, and Singapore rose from 3% to 10% of the overall research and development spending by those same companies. In the case of China, in the last decade, the research laboratories of foreign multinationals have burgeoned in number from 0 to 700 (UNCTAD 2005).

These phenomena have been accompanied by the first signals of a further shift of the barycentre of international research – historically located first in Europe and then, from the second half of last century onwards, in the USA – this time to Asia. The rapid economic growth of Asian countries like China is reflected by investments in research which combine traditional centralized planning with market mechanisms in technology-intensive sectors. This accounts for the sensation caused in recent years by the first manned missions of the Chinese space exploration programme and the purchase, in 2004, by the Chinese Lenovo company of one of the symbols of American technology, IBM computers.

More gradual, but no less significant, are the processes that have brought numerous research institutions, in countries like India, to the centre of international collaboration agreements. For decades an inexhaustible reservoir of recruits to European and American research institutions and Asian universities have in recent years begun to attract promising researchers from Europe and the USA. It has been estimated that currently almost one-third of the total number of graduates in advanced technical subjects originate from China, India, and Russia.

Often cited as a paradigmatic example of some of these changes is Yuan Longping, the Chinese geneticist known as the "father of hybrid rice", the owner of the biotech company which in 2000, the day after its quotation on the Shenzhen stock exchange, made its founder richer by the equivalent of around ten million euros (Pincock 2005).

2.7 The Eclipse of the Scientific Community?

Eric Engelhard is a Californian bioinformaticist. During the day, he works in a private company specialized in cancer research and which has patented more than 800 genes involved in certain forms of tumour. In his free time, he conducts his own research in a laboratory setup at home in a spare bedroom. With his specific skills, using equipment costing less than 500 dollars and sending the DNA samples collected to specialized firms that return the results via internet, Eric is trying to bioengineer a bee able to produce honey but without a venomous sting. He considers himself "an absolute partisan of research freedom" (Eudes 2002: 25) and it is ready to defend it against both the militant ecologists of his area and the federal laws that seek to hamper his experiments. But he is most opposed to what he regards as the damaging commercialization of research, which obstructs the free circulation of knowledge and forces researchers to abandon promising but financially unprofitable lines of inquiry. So Eric makes his results, methods, and software freely available on the web. With his colleague Katharine Nelson, who coordinated the Berkeley group within the international human genome mapping project before moving to the biotechnology company, he has founded the Central Valley Bioinformatics Interest Group, which has rapidly acquired around 200 members: "biohackers", as they have been dubbed to highlight their eccentric mix of technical skills, intellectual concerns, and affinities with computer hackers (Eudes 2002).

The example of Eric and the biohackers is certainly not to be taken as representative. Yet it synthesizes numerous dimensions at the core of current changes in technoscience and the same time problematizes a feature historically intrinsic to scientific practice: membership of a "community".

This concerns not just a change of codes and professional identity, or a transition comparable to that from little science to big science which transformed, according to the historian De Solla Price, "the Little Scientist [...] lone, long-haired genius, moldering in an attic or basement workshop" into "Big Scientist [...] honoured in Washington." (Price 1963:3). The difference is that the contemporary post-academic scientist is honoured in Wall Street and Hollywood. These recent changes are all the more significant because they configure a highly diversified set of practices, values, and conceptions of the scientist's social role.

Whether one can still speak in post-academic science of a "scientific community" in the full sense of the term is much too broad a topic to address here (on this, see, e.g., Ziman 2002; Nowotny et al. 2001). All the more so because the expression has now entered the vocabulary of commentators and scholars who sometimes employ it simply to denote scientific experts. However, if we take the term in its more strictly sociological meaning – whereby a community is distinguished by great internal homogeneity and the shared endorsement of specific norms and values – its application to post-academic science seems difficult. What professional ethos, for example, could be identified as the "glue" holding it together? Certainly not the "glue" which, 60 years ago, Robert K. Merton identified for big science in "institutional imperatives" such as disinterestedness and communitarism, and according to which, the results of research pursued in the name of a general growth of knowledge, rather than individual benefits, acquire value only to the extent that they are shared with other colleagues and with society as a whole. And this is not because today's scientists are less morally sound than those of the past. It is not difficult to document in the past, as well, severe breaches of Merton's norms in individual behaviours which did not cancel their functionality any more than a theft cancels the value of private property. Criticism and sanctioning by the scientific community of deviant behaviours asserted the value and functionality of the aforesaid norms. The point is that post-academic science institutionalizes and indeed advocates (as in the case of patents or increasingly marked overlaps between research and its commercial exploitation) practices contrary to a principle – that of communitarism – underpinning the notion of scientific community itself, and according to which "Research results do not count as scientific unless they are reported, disseminated, shared, and eventually transformed into communal property, by being formally *published*." (Ziman 2000:110).

The conventional opposition between individual interest and the public ownership of knowledge appears to break down: "I can do good science and make money" was the comment of a molecular biologist interviewed in the first years of this transition (Etzkowitz and Webster 1995). "There is nothing wrong if researchers claim intellectual ownership and want to register the fruits of their labours," the head of one of the leading European research centres in oncology said: "the exploitation of patents generates new funds for research institutes." (Dell'Oste 2005).

Nor does the transition to a post-academic science move in only one direction, and this further complicates identification of a new shared ethos. The pressures for the privatization of research results now assailing the pillar of communitarism are counteracted by other pressures for the more open dissemination and sharing of results and technological innovations, and which undermine the no less crucial principle of individual paternity, and therefore recognition. In the Mertonian ethos, individual paternity provided the necessary balance against communitarism: by pursuing knowledge for the sake of society, the scientist-obtained resources, prestige, and in some cases, the privilege of giving his or her name to natural objects or phenomena (the Golgi corpuscle, the Doppler effect, and Creutzfeldt-Jacob's disease).

As we have seen, post-academic science is anything but monolithic. It comprises movements and subcultures that sometimes assume the features of outright "countercultures", as in the case of biohacking or the movement for open access to publications, especially in its initial "subversive" phase. These subcultures often embody very different, if not contradictory, views of the role of researchers and therefore of their professional ethos. Moreover, membership of these subcultures seems somewhat fluid: the same scientist can cheerfully switch, as the bioinformaticist Eric Engelhard does, from a profit-oriented research organization to one which rejects copyright, or to the hacker culture. Patients' associations such as the Baschirotto Foundation for rare genetic diseases, which have developed to the extent that they now run their own research centres, do not disdain business agreements with private companies to register and commercialize patents with which to fund their search and care activities.[11]

From this point of view, the characterization of post-academic science as the mirror image of academic science is simplistic. In Ziman's terms, in fact, we should instead conclude that post-academic science is both private and public, concentrated on local problems but embedded in global networks, commissioned to solve practical problems but at the same time rather idealistic in its quest for knowledge.

This melange of norms and values not only contradicts the view of scientific communities as internally cohesive but also highlights their permeability. A specific scientific subculture may in fact be cultivated in close interaction with movements and normative and organizational cultures in the broader social context – industrial districts or firms' clusters, environmentalist associations, patients' groups, and mass media – with inevitable processes of what organization scholars have called "institutional isomorphism": namely, the tendency to assimilate practices and institutional models typical of one's interlocutors (Di Maggio and Powell 1983).

Overall, therefore, the main difference is not the presence of factors such as commercial interests, but their explicit embodiment at the identitarian and institutional level, and moreover, the capacity of these factors to structure the dynamics of research. If, for instance, in big science, the competition for resources such as prestige or funds was largely among specialists, to be then flatly transferred to the outside – receiving in return credibility with political power, commissions from business, and media interest – in post-academic science the competition for resources impacts on scientific practice itself.

Finally, the globalization of research induces reflection on how the professional ethos identified by Merton for big science was profoundly rooted in Western culture and tradition. According to numerous studies, the institutional development of modern science was driven by values typical of Protestantism (diligence in the empirical and individualized study of nature as manifesting the greatness of God; concrete commitment to practical activities as indicative of personal salvation) and of capitalist individualism (systematicity, methodicalness, and rationalism) (*see* Merton 1938a; Barnes 1985).[12]

---

[11] http://www.birdfoundation.org

[12] In general, according to Max Weber's well-known statement, "The belief in the value of scientific truth is the product of certain cultures and is not a product of man's original nature." (Weber 1922:110).

The ability of this ethos to act as an undisputed "glue" for post-academic science does not appear at all obvious, now that the industrialized West has ceased being the dominant reference scenario for technoscientific processes.

2.8 ... In the Meantime, Society Does Not Stand By and Watch

The belief in a society hostile to technoscience is also fuelled by a static vision of society counterposed against a dynamic vision of technoscience. On this view, technoscience constantly puts forward new proposals which society criticises and rejects. But it is a view which adopts an overly restrictive account of the interactions between technoscience and society – the transfer of information and knowledge from one to the other – and is therefore forced by its own premises to envisage an impermeable social context.

It is nevertheless evident that the transformations described here as configuring a post-academic science cannot have come about in a social vacuum. On the contrary, large part of these transformations are embedded within broader processes of economic, political, and social change whose description and analysis would be beyond the scope of this book. But by way of example, it is possible to foresee the decline of a model of research centred on the priorities and the regulatory capacity of government policies, whose apogee came with the period of big science, owing to the profound transformations wrought to nation-states by localist pressures and the transfer of powers to supra-national political institutions. Likewise, the relationship of research with business is part of transition to a post-industrial economy, characterized by flexibility, outsourcing, and ad hoc project-based consultation and research networks (Ziman 2000).

Some scholars have identified transformative processes shared by both the dynamics marking the advent of contemporary technoscience and broader social scenarios (Nowotny et al. 2001). The most prominent of these processes are the following:

(a) the growing uncertainty endemic to both knowledge production and more general individual and collective decision making, as in the "risk society" theorized by other scholars (Beck 1986);
(b) the increasing pervasiveness of an economic rationality which functions inter alia to filter the aforesaid uncertainty;
(c) the redefinition of the temporal dimension by expectations, forecasts, and scenarios, and by communication technologies centred on simultaneity which convert the future into an "extended present" [on this, *see* the clear treatment in Gleick (1999)];
(d) the redefinition of the spatial dimension and the compression of distance due to the growth of long-distance transport and above all to the development of communication technologies (Nowotrny et al. 2001:30–49).

Without entering into detail on each of these macro-transformations, what is to be stressed is the dynamic and co-evolving nature of the processes now redefining technoscience and society. The configuration that we have defined "post-academic science" takes place within, and at the same time reinforces, broader social, political, economic, and cultural changes.

The consequence is that when the technocrats condemn specific features of the social setup as resisting technoscientific advances, they forget that it may have been technoscientific advances themselves that have created such resistance. Consider, for example, how scientific and technological developments in numerous areas, from pharmacology to cosmetic surgery, to telecommunications, have contributed to people's perception that they can control their own destinies – a perception presumably not extraneous to the widespread "risk aversion" among non-experts[13] often cited as one of the main obstacles against the introduction of innovations such as GMOs.

It is within these scenarios of change in the role of technoscience, within broader social changes, that one should view phenomena like the increasing mobilization of citizens on issues concerning research and innovation, and the corresponding mobilization of researchers in the public arena. These phenomena are examined in Chap. 3.

[13] *See*, e.g., "We Play Too Safe", in *New Scientist*, 27 August 2005, 4.

Chapter 3
Citizens Enter the Laboratory Whilst Scientists Take to the Streets

When everybody is a superhero then nobody will be

(The Incredibles)

3.1 From Two Stubborn Parents to Seven Thousand Square Metres of Laboratory

France, the mid-1950s: The de Keppers' son was dying from a rare form of muscular dystrophy. Specialists and pharmaceutical companies were loath to concern themselves with this pathology: the former because they found it professionally frustrating to treat, since the likelihood of success was practically nil; the latter because the research and development of drugs would require huge investments to the benefit of only an extremely small number of patients. Dystrophy, indeed, used to be an "orphan" disease ignored by research, health institutions, and society.

Yet the de Keppers refused to watch their child die without doing anything about it. They contacted families in the same situation and worked with them to compile detailed accounts of the pathology's symptoms and evolution. They exchanged practical tips on how to alleviate the sufferings of their children; they signalled small glimmers of hope raised by articles published in specialist journals; and they circulated increasingly detailed information on the benefits and side effects of therapies.

On the death of their son in 1958, the de Keppers founded the *Association Française contre les Myopathies* (AFM) together with other parents who had suffered the same tragedy. The association promoted the systematic collection of clinical data and systematic trials on patients, and it also set up a data bank containing genetic information which might aid understanding of neuromuscular diseases and provide specialists with the basis on which to devise targeted research projects. Thereafter, the association carried forward a large-scale campaign to raise public awareness of muscular dystrophy which culminated in the organization, in 1987, of a 30-hour telethon to raise research funds.

Today the AFM has its own research laboratories employing the best researchers in the sector, and it has formed partnerships with the most prestigious international

M. Bucchi, *Beyond Technocracy: Science, Politics and Citizens*,
DOI: 10.1007/978-0-387-89522-2_3,

research institutes. The AFM has created the Genethon Institute, with 7,000 m^2 of laboratories, 162 employees, and an annual budget of 17.3 million euro – 85% furnished by the AFM and 15% by the Harvard Medical School – and the Institute of Myology, with 3,500 m^2 of laboratory space and over two million euros of funding per year. It has created its own "tissue bank for research", which since 1993 has collected 9,600 samples on 130 pathologies, providing the association's researchers with materials for more than 60 specialist peer-reviewed publications. It maintains 14 DNA data banks in Europe and Africa. More than 180 genes have been identified by the AFM researchers, or by projects supported by the association, which enhance understanding of 746 pathologies. In 2004, the AFM Telethon collected 104.7 million euros for research and therapy.

Similar associations, or ones explicitly linked with the AFM, now exist in numerous countries and also in Italy, where three research institutes and an institute of applied technology have been created using the Telethon receipts, and numerous research bursaries and projects are financed by funds which, from the telethon alone, amounted to more than 26 million euros in 2005. In the more specific area of rare genetic diseases, since 1989, the Fondazione Baschirotto has funded international research groups and has recently opened its own institute employing around 20 full-time researchers.

How can phenomena like the AFM be interpreted in the terms of a "clash of civilizations" between science and society? In face of the virtuous interaction that in this and other cases arises between experts and ordinary citizens, how is possible to view society – according to the technocratic scheme of things – as resisting and obstructing the progress of technoscience? Is society therefore obtuse, but only so on alternate days? Or does not the AFM show that oppositions and schematisms of this kind – the experts on the one side, and the ignorant and anti-scientific public on the other – are entirely unable to grasp either the dynamics that have brought science into society and the everyday lives of citizens, or the dynamics which in the last decade have brought society and citizens into the domain of science.

Public mobilization to support research, which as we have seen in the case of the AFM is not restricted to "external" support but engages with the very core of technical expertise, is part a broader phenomenon which further complicates attempts to reduce the challenges of technoscience to traditional technocratic schemes. Non-experts demand, with increasing success, to participate, to get involved, to have a voice in questions concerning science and technology. Levels of communication and social actors traditionally external to the sphere of research are able, in particular circumstances, to play a role in the definition and accreditation of scientific knowledge (Irwin and Wynne 1996; Bucchi 1998a). However this factor is assessed, it is bound to have a significant impact not only on the social role of science but also on the processes by which scientific knowledge is produced.

Of course, because this is an emerging phenomenon with still indistinct features, it is not easy to define precisely what is meant by "public participation in technoscience". The difficulty is compounded by the fact that the issue has almost simultaneously emerged as terrain for social mobilization, policy initiatives, and scholarly analysis. In what follows, the term "public participation" will denote the diversified set of situations and activities, more or less spontaneous, organized and structured, whereby non-experts become involved in, and make their own contributions to, agenda setting, decision making, policy formation, and knowledge and innovation production processes in the technoscientific field (Callon et al. 2001; Rowe and Frewer 2005; Bucchi and Neresini 2008).

3.2 Childhood Leukaemia in Woburn: "Hybrid Forums" and the Co-production of Knowledge

In the early 1980s, the inhabitants of Woburn, Massachusetts, were worried about what they thought was an unusually high incidence of leukaemia among their children. They began to meet in the evenings to talk about their concerns. One of them discovered that industrial companies were dumping polluting chemicals on sites not far from the town's residential area. The health authorities were called in, but the officials and experts reassured the families that there was nothing to worry about. The families insisted, however; they documented the potential hazards of the chemicals dumped in the area; they continued to collect information on the symptoms of the children falling sick; and they hired experts at their own expense, filed lawsuits, and organized public debates. They eventually managed to have the case re-opened by the authorities. When scientists from the Massachusetts Institute of Technology arrived in the town, they were given a dossier containing the results collected through the continuous monitoring of cases of leukaemia and other tumours over the previous 5 years. The research programme which ensued from that dossier led to the discovery of the "trichloroethylene syndrome", a disease (caused by one of the polluting chemicals dumped near Woburn) which damages the immune, cardiovascular, and neurological system, and which was subsequently identified in other areas of the USA (Brown and Mikkelsen 1990; Callon et al. 2001).

In cases like those of Woburn or the AFM, it is certainly not possible to view the public as an inert mass to be converted to the technoscience mission. Nor can we update the technocratic fable in the manner of some of its supporters, who patronisingly grant that, although the public does not know what it is talking about, it should be listened to, if nothing else because of some abstract democratic principle. A different matter is at issue: in the Woburn and AFM cases, the "local" knowledge of patients and their family members was not an obstacle against efficacious communication, as in the technocratic version of the deficit, nor was it a veneer of colourful folk lore to make expert knowledge more "politically correct". These, according to some scholars, are examples of that *co-production* whereby the role of non-experts has been essential for the production itself of technoscience. Expert and local knowledge were not independently produced in separate contexts and then happened to encounter each other; both resulted from shared processes which came about in "hybrid forums" in which experts and non-experts were able to interact (Callon et al. 2001).

One area in which this co-production has been particularly visible is that of biomedical research, where patient organizations have become increasingly active in shaping the research agenda. Particularly well-known and carefully studied is the case of AIDS research, where methods to test the effectiveness of drugs, and the term itself chosen to denote the disease [which was changed from the initial GRID (Gay-Related Immunodeficiency Disease) under pressure by American homosexual associations], were negotiated with activists and patients associations (Grmck 1989; Epstein 1996). In the mid-1980s, AIDS patients participating in clinical trials of the AZT drug (then considered a likely cure for the disease) developed a marked ability to contribute to and influence the experimental procedure, for example, by learning to recognize placebos and refusing to take them, thereby accelerating approval of the drug by the Food and Drug Administration. The trialling of another drug, aerosol pentamidine used to treat an AIDS-related disease, pneumocystis carinii pneumonia, was conducted in first person by groups of activists after scientists had refused to do so. In 1989, the use of aerosol pentamidine was approved by the Food and Drug Administration, which for the first time in its history authorized marketing of a drug solely on the basis of data collected by means of community-based experimentation (Epstein 1995).

In Italy, in 1998, following protests by groups of patients and their family members, oncologists found themselves trialling an anti-cancer therapy, the so-called "Di Bella method", which they had always previously refused to consider (*see* Bucchi 1998b).

But the already-cited case of the AFM is even more emblematic: here the families of patients did not merely apply pressure on science and policy but took direct action by collecting clinical data, producing – with the foundation of Genethon – a systematic mapping of genes implicated in myopathies hitherto considered impossible by the leading exponents of government research and exerting profound influence on the development of research in this sector. Many of the researchers working for AFM are at the same time geneticists and paediatricians, and thus combine research with daily therapeutic experience.

Associations of this kind frequently become significant gatekeepers for the circulation of technoscientific knowledge. In 1997, on reading a specialist article by a researcher at the University of Geneva, the married couple who founded the Associazione Baschirotto identified the same symptoms of the rare disease that had caused the death of their son. In the space of a few days, they were able to send the researcher biological specimens from their son and other patients with whom they were in contact as well as funding, thereby contributing to the development of research. Through other projects and funding programmes, the association has accelerated, and sometimes even forced, collaborations and exchanges of material among laboratories which otherwise would have been hampered by the dynamics of scientific competition. "Just think that a mouse modified to study metachromatic dystrophy took 6 months to arrive in Milan," said the founder of the association, Giuseppe Baschirotto. "By threatening to withdraw our funding, we forced the laboratories to collaborate because it didn't make sense to spend more money on producing a second mouse when there was one already available."[1]

Their differences notwithstanding, forms of involvement by non-experts are increasingly manifest in the now frequent cases of public mobilization on

[1]Original interview, 19 August 2005.

technoscience issues. Especially since the second half of the 1990s, in more- or less-organized forms, citizens have demanded closer involvement in decisions concerning the development of research and innovation. Most active in this field have been the "new social movements" and NGOs. These two phenomena are in fact closely interconnected: NGOs have provided crucial organizational support for new social movements and have been their main channel of recruitment, whilst the latter have given the former visibility and enabled them to intervene more effectively in decision making (Della Porta et al. 1999; Diani 1995; Della Porta and Turrow 2004). Nor should one overlook the local forms of protest usually centred on protecting health and the environment and opposed to the siting of installations which they deem dangerous in their area (waste disposal facilities, electricity power lines, cell phone antennas, and power plants). Although it is not yet clear whether these forms of mobilization pertain to the new social movements, they express an evident public demand for involvement in issues with high technical-scientific content.

Relations between new social movements and science have always been characterised by a marked ambivalence. According to social movement theorists, the distinctive features of such movements are the ways in which they construct an individual and collective identity, define the adversary, and structure a vision of the world put forward as an alternative to the dominant one (Touraine 1978, 1985; Meluccci 1989, 1996; Castells 1997).

It is evident that science and technology are bound up with each of these three features. First, science and technology are often an integral part of the "enemy" against which the new social movements mobilize. Second, they are viewed as instruments of the dominant power and as responsible for the perverse effects of globalization, especially now so that the connection between scientific research and economic interests is increasingly alleged (*see* Chap. 2; Etzkowitz 1990; Funtowicz and Ravetz 1993; Ziman 2000). Finally, once again the case of biotechnologies is paradigmatic: the science funded by, and therefore subservient to, the multinationals threatens the future of the environment (destruction of biodiversity), jeopardizes human health (harmful emissions), and increases the third world's dependence on the industrialized countries (by eroding the social bases of small-scale farming in the developing countries). For these reasons, it is an adversary to be combated.

Yet science and technology are also resources for the identity, organization, and action of the new movements themselves. In fact, the critique of the current model of development and the dominant economic paradigm is based on data concerning the depletion of environmental and social resources furnished by scientific analyses and forecasts (Moore 1995; Yearley 1995). Moreover, not only do the new movements rely heavily on the latest communication technologies (e.g., Internet and mobile phones) to organize their activities but they exploit more traditional media to gain access to the public arena and to exert political pressure. This "ambiguous, deep connection with science and technology" (Castells 1997:123; Yearley 1992) enables the new social movements, especially the environmentalist ones, to play a significant part in the production of scientific knowledge itself.

This participation may take place at various levels. First, a number of NGOs have set up laboratories and research facilities in order to have their own scientists produce independent scientific research (Yearley 1995). Also to be mentioned are the "science shops" set up by universities or networks of NGOs so that associations and citizens groups can commission research from those universities, or from other research bodies, at prices lower than market ones. Launched by Dutch universities during the 1970s, the science shop system has been introduced in various other countries in Europe, North America, and the Far East, where the initial intent to "re-orient science towards social needs" has been declined in various ways. In the case of the more mature Dutch experience, for example, "science shops started out as a counterculture phenomenon in the 1970s, but by the end of 1980s, most had become regular elements of university organization" (Wachelder 2003:253–4). Moreover, science shops can be viewed as an interpretation of the composite phenomenon termed "community-based research" in the English-speaking countries, and which comprises public participation not only in specific research projects but also more generally in research policies (Sclove 1998). An example particularly worth mentioning, and also interesting because it concerns a national context not traditionally conducive to civic mobilization, is that of "Japanese Citizens for Science". This is an association entirely supported by its membership fees. It produces a newsletter for its members and organizes seminars with researchers on topical technoscience issues, as well as commissioning technical surveys from independent experts, as it did, for example, in the case of emissions from the transmitters used by Japanese public television, which were found to exceed the levels permitted by law.[2]

The new social movements and NGOs participate directly in the production of scientific knowledge also by orienting research work in accordance with their beliefs (regarding experiments on animals, e.g.). They do so by seeking to disseminate certain theories (one of the best-known being the Gaia Theory) and to condition decisions on research policy (Yearley 1995:469–77). Finally, the new social movements have on occasion put themselves forward as the champions of a "true science", urging what they regard as a scientific community in league with political and economic power to regain its neutrality and independence – "the science of life versus life under science" (Castells 1997:127).

3.3 Technoscience Debated in the Courts

Jason Daubert and Eric Schuller were born with malformations due, they alleged, to Benedictin, an anti-nausea drug which their mother had ingested during pregnancy. Approved in 1956, Benedictin had been widely used until the 1970s, even though the clinical literature had already reported a possible link between the drug

[2] http://www.csij.org

and serious birth defects. Between 1977 and 1992, more than 2,000 patients sued Merrell Dow Pharmaceuticals, Inc., but very few of the lawsuits ended up in court. One of them that did, however, was the suit brought by Daubert and Schuller and supported by the opinion of eight experts based on in vitro and in vivo studies, pharmacological analyses showing the structural similarity of Benedictin with other teratogens, and a meta-analysis of various epidemiological studies. The district court deemed this evidence insufficient to bring the case before a jury, given that the more than 30 epidemiological studies conducted by the pharmaceuticals company had shown no significant link between the drug and the malformations in foetuses. On appeal, the court ruled that the data and the epidemiological findings submitted by the plaintiffs were inadmissible because they had not been published in scientific journals nor subjected to peer review. Thus, the Supreme Court had to adjudicate, not the plausibility of the link between Benedictin and Daubert and Schuller's birth defects, but the criteria that the courts should adopt in determining "if scientific evidence is admissible and sufficient to show causation" (Solomon and Hackett 1996:135). The standard hitherto used by numerous American judges had been the "Frye rule", so-called since the Frye v. US case in which the court had ruled that evidence obtained from a test conducted using a precursor of the polygraph was inadmissible. It reached its decision on the grounds that the methodological and theoretical bases of scientific evidence must be "sufficiently established to have gained general acceptance in the particular field in which it belongs".[3] In response to the decision, numerous memos were submitted, in support of both the plaintiff and the defendant, by scientists, associations, and scientific academies (among them the National Academy of Sciences and the American Association for the Advancement of Science), science publications (*New England Journal of Medicine*), lawyers, jurists, and companies.

The US Supreme Court's decision in the Daubert case, whilst acknowledging the importance of peer review by the expert community, re-affirmed the central role of judges in evaluating a certain piece of information as "scientific evidence" in a trial (Solomon and Hackett 1996).

As a result of judicial decisions such as in the Daubert case, a third important area of interaction between expert and lay knowledge has arisen in the area of law. At the strictly operation level, there are increasingly frequent cases in which the application of scientific methods and results in the justice system affect the dynamics of knowledge production. In 1991, a researcher complained that he had been subject to strong pressure by an official at the American Department of Justice to withdraw from publication a specialist article showing that the DNA tests used to identify the perpetrators of crimes were unreliable. According to the official, the article would have furnished lawyers with dangerous arguments with which to delegitimate the now widespread use of such tests (Roberts 1991).

More generally, over the last decade, a technicist view of the application of scientific knowledge in legal practice has been replaced by one where the law not

[3]Frye v. United States (1923), cit. in Solomon and Hackett (1996:135).

only utilises the tools of scientific research but actively participates in it, for example, by defining what can be patented as a scientific discovery, who can be considered a scientific expert, or even what counts as "scientific proof" (Jasanoff 1995; Mackenzie 1993) Indeed, in the "Kitzmiller v. Dover Area School District" case, initiated in September 2005 in Harrisburg, Pennsylvania, the court was required to decide whether the "intelligent design" argument was a scientific theory or a religious belief.

The sociologists of science and the jurists who have analysed this process describe the settings where laws are devised, and especially where they find interpretation by the courts, as ones of co-production between science and law in a context characterized by "an erosion of faith in legal processes and institutions", and by "the public's often expressed distrust of technical experts and their undemocratic authority" (Jasanoff 1995:4). Thus, courtrooms are used for experiments to solve the problem of arriving at socially endorsed decisions without delegating to experts-whether judicial or scientific - the task of unequivocally establishing what is to be taken as fact, on what bases and by what procedures. Legislation is being enacted that enables citizens to object to the standard neurological criteria used to establish death cerebral in the case of the New Jersey Health Statute - in case it conflicts with individual religious beliefs, thus making room for a sort of 'pluralist' definition of it (New Jersey Statutes, 1991; Tallacchini, 2002). The concept of democracy here acquires a meaning quite different from the traditional one: no longer a majoritarian decision-making process driven by hegemonic expert knowledge but rather the scrutiny of range of options using a procedure that gives a decision greater transparency without hampering its effectiveness.

Hence, the law as practiced in the U.S. courts has contributed significantly to the construction of a civic culture of science, both by revealing how the opinions of experts differ and by evincing "their underlying normative and social commitments in ways that permit intelligent evaluation by lay persons" (Jasanoff, 1995:215).

3.4 From Users to Innovators: How a Windsurfer Kept Himself Afloat and Became Something of a Designer

In the mid-1970s, the windsurfing enthusiasts gathered in Hawaii for the first international championships started to perform increasingly difficult jumps and flips. Many of these acrobatics caused injuries because it was impossible to keep hold of the board during the jump. For this reason, windsurfing remained a sport for a few reckless devotees. Then one of them, Larry Stanley, had the idea of attaching footstraps to the board: in this way, the windsurfer could jump without losing the board, and he or she could even change direction in mid-air. By 1998, there were more than one million regular windsurfers, and most of the boards sold around the world were fitted with the footstraps introduced by Stanley (von Hippel 2005).

This example illustrates yet another area – that of technology – where the involvement of different actors in knowledge creation is of increasing importance. Technology also warrants specific discussion because it is often the background for the other forms of interaction described here. Consider, for instance, the importance acquired by the new media in collective mobilizations, or the role of the internet in facilitating the gathering and exchange of information among patients and the families of patients suffering from rare diseases (*see* Chap. 5, Sect. 5). In the early 1990s, the members of the Amazonian Kayapo tribe made videos with a cheap portable digital camera to mobilize the inhabitants of villages scattered across a broad area against the construction of a hydroelectric dam. The tribe then made the case known to international public opinion by handing the videos to the mass media, and in this way eventually managed to get the project halted (Turner 1992).

Numerous studies have shown the various ways in which users are involved not only in the implementation of technologies but also in their design, and in developing the knowledge that makes them possible. In fact, artefacts can be re-interpreted, adapted, and in certain cases actually reinvented by users; their needs and point of views can be incorporated in the design process itself (Kent 2003; Pinch and Oudshoorn 2003; Eglash et al. 2004). Thus, a vinyl disc designed to re-produce a specific musical content can be used – as in the case of the hip-hop music invented in juvenile Afro–American communities – to produce entirely different sounds by means of "scratching" (Goldberg 2004).

Especially in cases where outright user-innovation communities come into being – for example, the free and open source software movement – participation by users in innovation processes goes well beyond the adaptation of initial projects to the needs of the final users, once again giving rise to the co-production of new knowledge (von Hippel 2005). Also of increasing importance are organizations set up both to protect consumers and to influence innovation processes so that they more closely reflect their needs. Manufacturing companies now make much use of informal sources (such as the discussion groups and virtual communities which arise around a technological artefact), and even monitor illegal practices like the downloading and exchange of files, to obtain information on the strengths and weaknesses of their products and how to improve them. In some cases, the companies have incorporated variants and solutions devised by the users.

It is difficult to tell whether these processes are effectively giving rise to what some scholars have called the "democratization of technological innovation" (von Hippel 2005). However, it is certain that, in technological field as well, discussion of innovation processes in contemporary society cannot be restricted to the conventional opposition between the forces that promote technology and the social dynamics that resist it, without taking account of how the distinction between innovators and users of technology grows increasingly blurred.

Less frequent, but not for this reason less interesting, is another "hybrid forum": collaboration between scientists and artists. In recent years, important institutions active in the field of scientific research like the Wellcome Trust, which also funds the Sanger Institute, one of the most important life sciences laboratories in Europe, have launched schemes to encourage collaboration between researchers and artists. The initial aim was to exploit various forms of artistic expression (theatre, dance, performances, and figurative arts) in order to reach the general public and interest it in science. Over time, however, these schemes have been used as ways to develop the creativity of both artists and scientists, giving the former stimuli and materials from science, and the latter opportunities to explore new aspects of their research through interaction with the arts. An example is provided by a project on visual agnosia – a pathology which distorts vision following damage to the cerebral cortex – funded by the Wellcome Trust's Sci-Art scheme. The videos produced by the artist collaborating on the scheme enabled the researcher concerned to develop new tools for the diagnosis of visual agnosia.[4] Numerous research institutes, from NASA to the Xerox Parc of Palo Alto, have "artists in residence" who, by working in close contact with scientists, are able to re-interpret the institute's research in their own language, thereby stimulating the researchers to reflect upon their work from a different perspective.[5]

The principal types of "hybrid forum" described thus far – patients associations, new social movements, the law, art, and technology – are obviously not to be taken as "pure" types clearly distinct from each other, but rather as contexts which frequently interact in the co-production of technoscience. Mention has already been made of how technology is not only a terrain of interaction per se but also furnishes devices which enable and intensify interactions among non-experts and between the latter and experts. Consider how the associations and informal networks of patients affected by the same pathology make massive use of the Internet to exchange information and advice. But the point applies to the other forums as well, with the knowledge co-production process increasingly becoming more substantial by moving from one forum to another. This is the case of a collective mobilization which leads to a court case or vice versa, or of a protest which gives rise to an association of patients or citizens able to commission research or surveys.

It is also under the increasing impact of the hybrid forums and processes described here that the attention of researchers and operators has concentrated on a variety of initiatives expressly aimed at promoting the involvement of citizens in issues concerning science and technology.

3.5 Everyone Around a Table: Promoting Civic Participation in Technoscience

In November 1996, a rather unusual meeting was held in Copenhagen. It was attended by 60 experts, 60 politicians, and 60 citizens. The problem to be discussed was a serious one: for years, the quality of the water extracted from Denmark's aquifers had been deteriorating, and in some places was no longer drinkable. It was widely believed that the origin of the problem was the use of pesticides and fertilizers in

---

[4] http://www.wellcome.ac.uk/sciart

[5] A partial list of these institutions is available at http://www.artistinresidence.org. These, of course, are initiatives that have impacts in terms of the public visibility of the research institutes concerned: all the international news media, for example, reported that NASA had chosen the well-known musician and multimedia artist Laurie Anderson as its first "artist in residence".

agriculture, but so far discussion between the disputing parties had not been constructive. The environmental associations believed that the only solution was a switch to organic farming, with the banning of all substances suspected of causing the water pollution.

The chemicals industry and the farmers argued that there was no unequivocal proof that the water was harmful. They also maintained that the current levels of pollution were due to products used in the past but no longer on the market. They therefore proposed a gradual reduction in the use of agricultural chemicals but emphasised the high economic costs of restrictive measures on pesticides and fertilizers. The water companies and the municipalities, for their part, wanted the quality of drinking water to be protected by reducing land exploitation and by giving local authorities greater powers in the management of water resources.

Before the meeting, every participant had received a dossier on drinking water and the aquifers. They then heard descriptions of five projects to solve the problem proposed by the Danish Council for Agriculture, the Danish AgroChemical Association, the Danish Association of County Councils, an environmental organization active in the protection of groundwater, and the Danish Water Supply Association. After every presentation, the participants could ask for clarifications or put questions. The five projects were then put to the vote: the winner was the one proposed by the Danish Water Supply Association, which received just a few votes more than the environmentalist action plan. The final report was delivered to Danish members of parliament and presented at a press conference.

This is the account of a "voting conference" organized by the Danish Board of Technology, an institution tasked with conducting participatory assessments of technology and, on this base, formulating advisory opinions for Parliament (Joss and Bellucci 2002).

Particularly, since the mid-1990s, local, national, and international public institutions as well as NGOs in many countries have devoted significant efforts to creating opportunities for citizen participation with regard to potentially controversial science and technology issues like GM food, genetic testing, transport technology, and ozone depletion.[6] Political institutions have also started to consider "citizen participation" as a necessary policy provision in the field of research and innovation, and with special regard to highly sensitive fields like biotechnologies, the siting of radioactive waste disposal facilities, or more in general, sustainable development.[7] In some countries, for instance, Switzerland, specific agencies have been established to undertake "participatory technology assessment" of upcoming innovations on behalf of parliaments or governments (Joss and Bellucci 2002).

Examples of the inclusion of citizens in decisions on technology can be drawn from a variety of contexts, even local, and especially in regard to health policies.

---

[6] See in this regard the broad debate on "deliberative democracy": for example, Bobbio (2002) and Pellizzoni (2004).

[7] *See*, for example, European Directive 2001/18/EC on release of GMOs into the environment or the "Agenda 21" document of the United Nations (http://www.un.org/esa/sustdev/documents/agenda21/index.htm).

In Catalonia, the regional health board for several years has been experimenting with ways to involve patients and their families in the definition of such potentially conflict-provoking matters as the criteria for establishing priority on hospital waiting lists. Through focus groups and other consultation procedures, the point of view of the doctors and the medical staff on the criteria for establishing priorities in cataract operations or hip replacements has been integrated with that of the patients and their relatives. Thus, a criterion more important for the experts – the relative severity of the pathology, for example – has been balanced with criteria such as the patient's self-sufficiency or ability to rely on family support for his/her everyday needs (Espallargues et al. 2005).

In the technological area, one of the first examples in Italy has been the 2004 experimental "consensus conference" organized in Lombardy. Two groups of around 15 citizens were selected on the basis of a series of criteria (gender, age, area of residence, education level, non-membership of environmental organizations, research institutes, and biotech companies) and brought together for an entire day to discuss whether open-field trials of GMO should be carried out in Lombardy. Before discussion began, the first group listened to presentations by science experts and "institutional" stakeholders (for instance, members of regional advisory committees). After an initial phase of free discussion, the second group was instead able to choose three "witnesses" from a list of scientific experts and stakeholders and interrogate them so that it could reach its own conclusions. The final document, which substantially approved open-field GMO trials provided guarantees were put in place, was sent to the regional institutions (Pellegrini 2004).[8]

The promotion of public participation in the area of science and technology is often justified by the sponsoring institutions in terms of enhanced citizenship and democratic participation. This rationale is sometimes expressed in the more sophisticated argument that advances in research and innovation are challenging the standard forms and procedures of democracy. Hence, new forums and opportunities are required in which complex technoscience issues can be addressed without sacrificing the needs of contemporary democracy. However, not infrequently, the sponsors more or less implicitly expect that opportunities for participation will forestall heated public controversies on sensitive issues related to science and technology and restore otherwise declining public trust in science. Indeed, a number of initiatives in the area of science and public participation, like the wide-ranging "GM nation" debate conducted in the UK in 2003, have been launched after significant public mobilization on a particular issue (Jasanoff 2004a). This expectation – in this or its even more cynical version where participation is seen simply as providing stronger public legitimation for decisions already taken elsewhere (Callon et al. 2001) – may in some cases be expressed so explicitly that participants

---

[8] The experimental consensus conference on GMOs was conducted for the Lombardy regional administration by the Giannino Bassetti Fondazione in collaboration with the Observa-Science in Society association and research centre.

and commentators begin to suspect that the institutions promoting public participation initiatives see them, to some extent, as the "prosecution by other means" of the missionary vocation that we have seen associated with the technocratic vision.

Sponsored initiatives for public participation in science have taken a variety of forms in terms of:

(a) the nature, number of participants, how they are selected, the time frame, and the geographic scale;
(b) the method by which the public input is gathered;
(c) the extent to which this input may be binding for policy decisions;
(d) the type of issue at stake (Rowe and Frewer 2000).

Participants may, for instance, be representatives of the stakeholders (i.e., the "litigants": companies, environmentalists, and consumers) as in the "negotiated rule making" exercise, or ordinary citizens selected (according to certain criteria) to represent the public as in the "consensus conference" (a model first experimented with in Denmark in the late 1980s). The number of participants may be fairly small, as in the case of citizens' juries, or quite large (public opinion surveys). Events may last a few minutes (public opinion surveys) or several months (public hearings, negotiated rule-making); the geographical scale may range from the very local to national and (more rarely) transnational. The methods used to obtain the public's input may be multiple choice questions, moderated or free discussion by participants, the questioning of expert witnesses, or presentations by agency representatives, expert witnesses, or stakeholders. Participant input may be strictly binding (as in the case of referenda) or simply offered to policymakers as additional support for their decisions (consensus conferences, citizens' juries), or it may even consist in no more than opinions which may well not be included in the final recommendations (public hearings).

The themes addressed may be very general topic (in 1996, a consensus conference was organized in Denmark on "The Consumption and Environment of the Future") or single issues (genetic testing, GM food, and cloning). The questions asked of participants may concern the implementing of specific decisions at the local level (e.g., choosing the most appropriate site for a new waste disposal facility) or the devising of broad, long-term scenarios (e.g., the future of transportation).

Table 3.1 sets out some of the most widespread forms of public participation elicited by a sponsor, that is, an organization – public body, research institute, private or non-profit enterprise – which convenes participants around a table to discuss a technoscience issue.

Is assembling citizens, experts, and stakeholders around a table a useful exercise? Put in these terms, the question is not only difficult to answer but also largely misleading. And it at least partly explains why evaluation of these initiatives has not yet provided clear indications. It is necessary, in fact, to specify who are involved in these initiatives and what objectives they are supposed to achieve. The concept itself of "effectiveness" can be articulated on several dimensions, as well as

Table 3.1 Some of the most widespread forms of public participation in technoscience elicited by a sponsor

Participation method	Nature of participants	Time scale/ duration	Characteristics/mechanism
Referenda	A significant proportion of national or local population	Vote cast at single point in time	Vote is usually choice of one of two options. All participants have equal influence. Final outcome is binding
Public hearings or inquiries	Interested citizens, limited in number	May last many weeks or months or even years	Presentations by agencies regarding plans in open forum. Public may voice opinions but have no direct impact on recommendation
Public opinion surveys	Large sample of population	Single event, lasting few minutes	Input gathered through a questionaire administered face to face, by telephone, via post, or e-mail
Negotiated rule making	Small number of representative of stakeholders groups	Uncertain; usually lasting days to months	Working committee of stakeholder representatives (and from sponsor). Consensus required on specific question
Consensus conference	Generally, 10–16 members of public, selected as representative	Preparatory demonstrations and lectures to inform panellists about topic, then 3-day confernce	Lay panel with independent facilitator questions expert witnesses chosen by stakeholder panel. Meetings open to wider public. Conclusions on key questions made via report or press conference
Citizens' jury/ panel	Generally, 12–20 members of public selected as representative	Generally involve meetings over a few days	Lay panel with independent facilitator questions expert witnesses chosen by stakeholder panel. Meetings not generally open. Conclusions on key questions made via report or press conference

Source: adapted from Rowe and Frewer (2000: 8–9)

from the different perspectives of the actors directly involved, of those who are in some way affected, or even of those excluded from the initiative. It is evident that if the expectation is to use institutionalized forms of participation to resolve every potential conflict and form of dissent, observers or the sponsors themselves may be disappointed when this objective is not achieved, as in the case, for instance, of the British "GM Nation" debate (*see* Chap. 1, Sect. 3). For a participant, on the other hand, the final outcome of an initiative may be of secondary importance with respect to the procedure itself. In the case of the consensus conference on the GMO trials in Lombardy, the mere fact of having been consulted was valued by the participants as a positive element in itself – also because it has hitherto been lacking – in decision making on such issues. Conversely, the perception by participants that the sponsors have distinct expectations that the initiative will yield a predetermined result – support for a particular policy decision or simply closure of the debate – may in many cases decisively prejudice their evaluation of the initiative and therefore its eventual "success" (Purdue 1999; Irwin 2001).

The criteria employed to assess participatory initiatives have referred both to their public "acceptability" (e.g., representativeness, independence, early involvement of the public, influence on policy decisions, and transparency) and to "process" considerations relative to their design and implementation (e.g., the availability to participants of the resources necessary for their task, clear definition of the task, structured decision making, and cost-effectiveness).

3.6 Science and Public Participation: A General Interpretative Framework

The proliferation and variety of participatory mechanisms and the problem of finding common definitions certainly reflects the *statu nascenti* instability of the field and is at least partially responsible for the difficulty of deciding "what works best and when": that is, of assessing the effectiveness of each specific technique. Nevertheless, since participatory initiatives first made their appearance, attempts have been made to categorize them on the basis of such dimensions as objectives, type of participants, and the extent to which the procedure is structured. A recent study draws up a typology of participatory mechanisms with a view to evaluating their effectiveness. The authors consider a general aim of "public engagement" to be "maximizing the relevant information flow (knowledge and/or opinions) from the maximum number of relevant sources and transferring this efficiently to the appropriate receivers" (Rowe and Frewer 2005: 263) (*see also* Rowe and Frewer 2004). Depending on where the emphasis in the process is placed, three broad categories of public engagement can be thus identified:

- *public communication*, "maximizing the relevant information flow from the sponsor [...] to the maximum number of relevant population",
- *public consultation*, "maximizing the relevant information flow from the maximum number of the relevant population and [...] transferring it to the sponsor", and
- truly *public participation*, "maximizing the relevant information from the maximum number of all relevant sources and transferring it [...] to the other parties".

Differences between specific participatory procedures can be related to a series of variables associated with the above objectives (e.g., maximization of relevant participants and maximization of relevant information from participants). This typology has several advantages: most notably, it highlights similarities and differences between mechanisms and thereby paves the way for conceptual clarification and thorough impact evaluation. For instance, consensus conferences, citizens' juries, and action planning workshops can be treated as a homogeneous cluster of participatory forms as they all involve a controlled selection of participants, facilitated elicitation, an open response mode, and unstructured group output (Rowe and Frewer 2005:281).

Nonetheless, the typology may not be fully satisfactory for a series of reasons. First, it anchors public engagement to a notion of information flow, described as a rather mechanical process of "transfer", which seems largely to reprise the limitations of the deficit model and traditional communication paradigms; the main difference being that it envisages the possibility of two-way transfer (i.e., not only from the sponsor/experts to the participants but also from participants to the sponsor/experts)[9]. However, hybrid forums often involve not only the exchange of information among the actors concerned but also the negotiation and production of new collective identities (Callon 1999).

Second, defining relevance as a key concept for the typology is only unproblematic if a specific point of view is adopted. Who defines which information is relevant? Who defines which population is relevant? Is it the sponsor promoting the specific participatory initiative? The potential participants? In the case of muscular dystrophy patient associations, the relevant groups did not exist until thorough interaction between them and the experts became possible; just like the disease, they became visible and relevant only through this interactive process (Callon 1999).

This brings out a third, and probably more substantial, shortcoming of the typology: that is, it is limited only to mechanisms actively promoted by a sponsor. In that I have instead adopted a broader definition of participation, I shall propose an interpretative framework able to account also for "spontaneous" participatory forms, for example those not deliberately elicited by a sponsor in the sense indicated above: public mobilization and protests, patient associations shaping the research and care agenda, and community-based research.

This framework is partly based on the one used by Callon to classify hybrid forums and adopts one of its key dimensions: the intensity of co-operation among different actors in knowledge production processes (Callon et al. 2001:175). Whilst intensity should of course be understood as a continuum, some key gradations can be identified along it: what Callon et al. correspond to "access points" where non-experts can intervene. One such point is the moment when laboratory results are "translated" to real-life situations, which is a crucial stage in the stabilisation of scientific knowledge (Callon et al. 2001:89ff). At that point, contradictions and conflicts may emerge between specialist and lay knowledge, with non-experts questioning the extent to which laboratory data can be applied to their own specific situation. This was, for example, the case of people living close to the Sellafield nuclear re-processing site, who used data collected by themselves to contradict the reassuring statistics of experts on the number of leukaemia cases in their area and eventually obtained an official enquiry, or the above-mentioned case of the Cumbria sheep farmers whose concrete experience of the peculiarity of Cumbrian soil gainsaid predictions based on expert models that the contamination would soon disappear (Wynne 1989).

A second and more substantial degree of participation corresponds to the access point offered by what has been called "the definition of the research collective":

[9] *See* Chap. 1; for a critique of a "transfer" model in the communication of science, *see* Bucchi (2004).

for instance, when members of AIDS patient associations manage to gain involvement in the design of experiments and drug trial tests, thereby broadening the research collective to include non-researchers.

The public may even participate in the initial recognition of research problems, for instance, by making a particular event or series of events leave the limbo of happenstance and enter the realm of problems warranting expert interest and attention. The public may also accumulate the initial stock of knowledge required to make professional research possible and worthwhile. For instance, in the 1980s, it was the action by Woburn County residents in gathering by themselves epidemiological data and information on a suspiciously high number of children leukaemia cases in their area that eventually persuaded MIT to initiate a research programme which discovered genetic mutations caused by trychlorethylene (Brown and Mikkelsen 1990). Similarly, the mobilization of patient associations like the French AFM has been crucial in prompting fruitful research on genetic diseases.

The second axis of our diagram plots the extent to which public participation is elicited by a sponsor: what could be defined, with a certain amount of simplification, and the degree of "spontaneity" of public participation. Here again, the variable should be viewed as a continuum, with the participatory initiatives described by Rowe and Frewer at the upper end of the axis and protest movements and research activities of patient and resident organizations at the lower end. Figure 3.1 gives a graphical re-presentation of the space defined by these two dimensions, together with some purely illustrative examples.

A wide variety of forms and cases of public participation can be mapped in this space. The upper-left quadrant comprises forms typically elicited by a sponsor and characterized by low-intensity participation by non-experts in knowledge production, for example, a public opinion survey. The lower-left quadrant contains spontaneous mobilizations which do not significantly impact on the dynamics of research, for example, residents' protests against the decision to locate a radioactive waste site in their area. The lower-right quadrant includes "spontaneous" forms of knowledge co-production, such as those exemplified by the Woburn residents or by the AFM. Finally, a participatory initiative like a consensus conference on a technoscience issue organized by a sponsoring institution can be placed in the upper-right quadrant (high degree of elicitation and high degree of intensity).

Over time, public participation with regard to a certain issue may move along one or both dimensions: for instance, when a public protest induces an institutional sponsor to organise a consensus conference or a citizen panel, or when patient families initially get together to lobby research institutions or drug companies and in the long run decide to establish their own research facilities.

The "open-endedness" of public participation is also emphasised in this interpretative framework. By "open-endedness" is meant that the outcome of public participation is rarely predictable on the basis of its structural features or on the basis of the sponsor's objectives: a public protest, for instance, may lead to re-negotiation of a consensual decision, just as a participatory initiative originally designed to produce a consensus document may bring to light and radicalize conflictual positions, both among actual participants and, especially when the

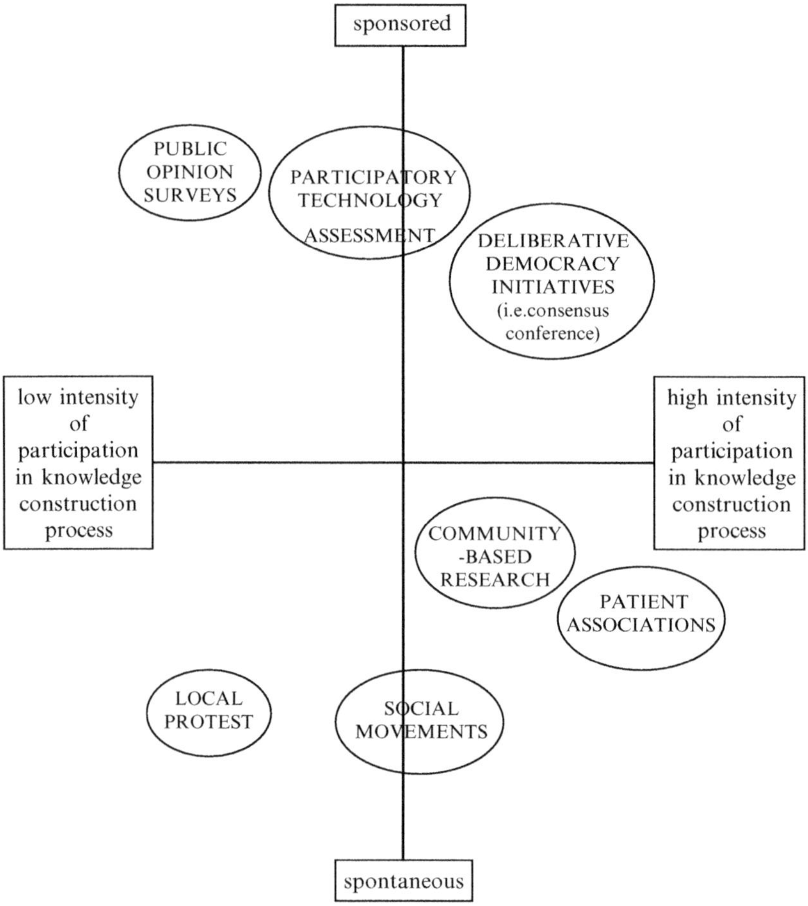

Fig. 3.1 A map of the forms of public participation in technoscience

conflicts are reported by the media, in the broader public arena. Some degree of apprehension for this open-endedness may be regarded as a key factor accounting for the sometimes evident temptation, on the part of research bodies and other institutions, to tame unruly public participation through formal initiatives.

Overall, it seems excessive to predict that the development of public participation in technoscience will lead in the long run to the disappearance of professional experts and their replacement by widespread and socially diluted knowledge. One reason for this is that the model of knowledge co-production, though undoubtedly commonplace in certain areas of biomedical and environmental research, does not seem equally applicable in other fields of scientific enquiry such as theoretical physics. Second, the use of inevitably generic labels like "non-experts" or "lay public" should not induce us to flatten the intrinsic variety of citizens' involvement and their significantly differentiated capability and interest to shape knowledge production processes. Indeed, some the most intense examples of participation actually involve highly motivated, highly informed groups – "quasi-experts" among non-experts, so to speak – that leave large parts of the public potentially disenfranchised. Accordingly, the interpretative framework proposed seeks to account for the simultaneous co-existence of different patterns of participation which may coalesce depending on specific conditions and on the issues at stake, from the "zero degree" of participation entailed by the deficit model to the most substantial forms of co-operation between experts and non-experts.

3.7 The "March of the Test-Tubes": Scientists Take to the Streets

Seen from a distance, the protest march of 12 February 2003 seems no different from the many demonstrations that take place in Rome's Piazza Montecitorio. The demonstrators waved their banners and chanted their slogans. But on closer inspection, there was something different about the protesters. They were holding (or wearing) white coats, and as they thronged the square, in a gesture of symbolic and polemical "restitution", they brandished test-tubes, microscopes, books, even a brain in a glass jar, and a pear ("There's also research in agriculture"). Among the demonstrators were well-known Italian scientists such as the astronomer Franco Pacini and the physicist Carlo Bernardini, who admitted "I don't like protesting, I don't like street demonstrations. But sometimes needs must." (*La Repubblica* 13 February 2003:23). At the last moment, the Nobel Prize-winner Rita Levi Montalcini decided not to take part in the demonstration, "perhaps because she was embarrassed by its more theatrical aspects" (*Il Corriere della Sera* 12 February 2003:13), although a few days earlier she had urged researchers to stage an outright "general strike". "If all researchers", she declared, "downed tools and refused to work, they could create a real problem. A mobilization of the kind has never happened before, but this might be the right moment for it to happen." (*La Repubblica* 2 February 2003:26). The protest was against the law reforming the National Research Council, the National Institute of Astrophysics, The National Institute of Materials Physics, and the Italian Space Agency, approved on 24 January by the Council of Ministers. The reform reorganized the National Research Council into a small number of departments, distributing the 108 existing centres among "macro-areas" of strategic research and incorporating other bodies, like the National Institute of Materials Physics.

The protest was not just a salient moment in a wide-ranging public debate on the problems of Italian scientific research. Rather, it was particularly significant stage of a phenomenon – the mobilization of researchers – unprecedented in Italy until only a few years ago. Hence, whilst the public has increasingly entered laboratories and research, the scientists have mobilized and taken to the streets.

A few years before, on 5 November 2000, the *Sole-24 Ore* newspaper had published a petition "for the freedom of research" signed by more than a thousand researchers, among them the Nobel prize-winner Renato Dulbecco, which criticised the actions of the then minister for agriculture and forestry policies, Alfonso Pecoraro Scanio.[10] The minister had made access to ministerial funding conditional upon an undertaking by researchers to discontinue the field trialling of GMOs. According to the petition's signatories, Pecoraro Scanio's provision was likely to exclude Italian research from one of the most promising sectors in the biological sciences and to frustrate the investments made in past. The provision added to a series of restrictive measures – among them already-mentioned decision to ban the marketing of derivatives from genetically modified maize and rapeseed (*see* Chap. 1, Sect. 4) – introduced by the Italian government and already criticised by leading representatives of the Italian research community.

The scientists' protest culminated in a huge public demonstration in mid-February 2001. On that occasion, representatives of researchers, among them Rita Levi Montalcini and the well-known pharmacologist Silvio Garattini, met members of the government and the opposition. Their main demands for research in Italy, explained to the biologist Edoardo Boncinelli during a television interview, were "money, meritocracy, and organization". Thereafter, there were further controversies (also reported by the international science journals) and renewed protest by researchers. In December 2002, the decision by agriculture minister Alemanno to suspend previously authorized open-field GMO trials provoked a fierce reaction and an open letter from a group of scientists to Prime Minister Berlusconi (*Il Corriere della Sera* 7 December 2002:16).[11] Similar protests were held in France, where in 2004 researchers mounted a large-scale mobilization against cuts in research funding, and analogous appeals and campaigns have become common in other countries as well.[12]

Without going into the issues that gave rise the protest is possible to put forward some considerations on the phenomenon which also relate to more general changes in the relationships among science, politics, and public opinion.

First, the large-scale public exposure of researchers happened at a time of increased public awareness of science and some of its applications. Episodes like BSE or the GMO controversy have been, at least to a certain extent, interpreted by the mass media and the public as examples of the scant capacity, indeed, of scientific experts to furnish appropriate solutions to society's needs. On the occasion of one of the protests, a television interviewer announced "This science is frightening", before going on to ask Levi Montalcini about the above-mentioned emergencies.

---

[10] http://www.ilsole24ore.com/cultura/liberta_ricerca/appello_0511.htm

[11] *See also* the Sunday supplement of *Il Sole-24 Ore* of 18 March 2001, which, at the height of an election campaign, published a series of articles – by Giovanni Bignami, Luca and Francesco Cavalli Sforza, and Cinzia Caporale – directing the attention of the contending political coalitions to the problems of Italian research. In September 2002, after an article published in *La Repubblica* disclosed the contents of the bill to reform research bodies being prepared by the Ministry of Universities and Research, a further petition by researchers was published on the web site of *Le Scienze* journal, http://www.lescienze.it

[12] *See*, for example, the appeal to the European Union "against the decline of science and the brain drain" signed by five Nobel Prize-winners and numerous other scientists and published by the European daily press in June 2002, *see La Stampa* (22 June 2002:25).

The second consideration concerns the way in which we have grown accustomed to perceiving street demonstrations, particularly, in terms of media coverage. In most cases, people do not today protest in the name of some ideal but rather to protect their interests: whether they are dock workers or cattle farmers makes no difference. But scientists are among the few categories which the public have always regarded as disinterested – as the advocates, not of a particular interest, but of the universal and superior value of knowledge. This is confirmed by the rare research conducted in this area: compared with the citizens of other European countries, Italians place great trust in science and scientists for precisely this reason. But the most recent studies on the public perception of science report that scientific activity is increasingly perceived as biased (*see* Chap. 1; Observa 2005). There is a consequent risk that the claims of scientists will be equated to the more usual ones of other actors and organizations, raising new doubts and concerns ("So now they want money as well?"; "And why don't they want to be regulated?"; "Have they got something to hide?").

Whilst the shell of the scientific community ensures an image, however idealized, of internal consensus, the latter easily breaks down in the public arena. In February 2001, the Green Party immediately organized street protests by "ecological" scientists, whilst the impact of the agreement reached with Environment Minister Pecoraro Scanio and Prime Minister Giuliano Amato was played down by a number of researchers.

Researchers also run the significant risk that their constant complaints about a lack of interest in their work among politicians and the hostility of public opinion may turn into a self-fulfilling prophecy. If they continue to insist at every opportunity on their isolation and weakness, it is highly like that these problems will be exacerbated.

Finally, and this is perhaps the most complex and significant consideration, science's "taking to the streets" is a strategy which is difficult to reverse. Overtly seeking public support requires being prepared to sustain the pressure of public opinion; it means being able to justify the priorities of research and their compatibility with those of society and politics; it requires, by way of example, being ready to devote, despite all misgivings, time and resources to trialling a therapy demanded *a furor di popolo*, as happened not so long ago in the Di Bella case, or indeed as happened in the USA, when the definition of AIDS was changed under pressure by activist groups (*see* Sect. 2; Epstein 1996).

It is difficult to determine whether this "exposure" is an advantage or a drawback. The fact that it occurs with increasing frequency suggests that it is bound to become an intrinsic feature of contemporary science. It should be stressed, however, that to date scientists had always been very careful to restrict it, thereby protecting the autonomy of science that they often invoke as the guarantee of its authentic freedom and the necessary condition for its conduct.

The mobilizations of researchers can also be interpreted in light of more general changes in the role of technoscience in contemporary society. First to be mentioned is the process which we may call the "politicization of technoscience". This expression obviously refers to the fact that public exposure of science increasingly takes place in the forms typical of collective mobilizations, such as the street demonstrations widely covered by the media. The problems of research policy and more generally the governance of technoscientific innovation seems, for various reasons, increasingly less manageable within what C. P. Snow termed "closed politics" (*see* Snow 1960; *see also* Chap. 1, Sect. 4) – cloistered negotiation between policymakers and scientific experts characterized by a more or less explicit trade-off between economic resources and the legitimation of policy decisions through expertise.

Among the reasons for this change, and among the meanings assumed by the politicization of science, is the increasing public perception and representation of the scientific community as divided and fragmented. Increasingly apparent is disagreement among researchers on issues of interest to public opinion and policy decision making, for example, the danger of GMO to human health or to the environment, or the best way to dispose of radioactive waste (as in the case of Scanzano Jonico, an Italian site chosen for radioactive waste disposal which provoked fierce controversies among scientists like that between the Nobel Prize-winner Carlo Rubbia and the president of the Institute of Geophysics, Boschi).[13] A first, emblematic, signal of this phenomenon was apparent in Italy some years ago with the Di Bella case, which, among other things, dramatically highlighted the scientific community's difficulty in finding a spokesperson able to represent it on the public stage (Bucchi 1998b).

Disciplinary boundaries turn into cleavages when it is necessary to define problems and who is competent to solve them. "I don't know him personally," said Rita Levi Montalcini concerning the appointment of Adriano De Maio as commissioner of the National Research Council, "and therefore I can't judge him. What I can say, however, is that he's an engineer and knows nothing about biology."[14]

Also specifically in regard to protest, the prevailing image is that of a scientific community unable to find a spokesperson and divided among the positions and "parties" typical of political campaigns. Thus, researchers in favour of the reform fostered by the Research Minister Moratti are called "the pro-Moratti front", "the presidents' alliance", or with even more blatant analogy to referenda, "the yes front", whilst their opponents are often dismissed as "lobbies". Indeed, politicians from the opposition equated the resignation of the President of the National Research Council, Bianco, to the "Piave line".[15] And Bianco himself urged researchers to rally to his side, complaining that "there is a section of the scientific community, also occupying responsible institutional posts, which does no more than raise doubts without expressing a clear and incisive position. I say to them that this is the attitude of losers and it will bring no benefits to Italian research."[16]

A significant part of the criticism of the reform by researchers, moreover, is not directed against its specific contents or consequences for research in the strict

---

[13] *See*, for example, F. Foresta Martin, *Corriere della Sera*, 27 November 2003.

[14] C. Di Giorgio, *la Repubblica*, 2 February 2003.

[15] C. Di Giorgio, *la Repubblica*, 2 February 2003. In 1918, during the First World War, Italian troops fought against the Austrian army a famous battle along the Piave river.

[16] C. Di Giorgio, *la Repubblica*, 14 May 2003.

sense, but against the decision-making procedures that have produced it. The reform is described as "anti-democratic", "imposed from above", and without sufficient consultation of the subjects involved.

Extremely frequent in mass media coverage of protests is the equating of research with other issues on the political agenda. When questioned about the possibility of opposition in parliament against the government's proposed legislation, Levi Montalcini exclaimed: "Not a chance! They voted for the Cirami law, they voted for the bill on false accounting, the one on rogatories. They'll vote for this one as well."[17] Levi Montalcini's declaration was immediately taken up and amplified by the political parties: "We'll take to the streets for research just as we did for justice," declared a representative of the Center left coaltion.[18] The wholly political nature of the initiative was underlined at the rally held during the demonstration by the president of the National Research Council himself: "I must thank Silvio Berlusconi. I haven't taken part in a demonstration since the one on Trento and Trieste question, and that was when I was thirteen years old."[19]

Moreover, the press does not identify opinion leaders solely on the basis of their roles within the scientific community; it also does so on the basis of their closeness to political leaders. This is the case of "the premier's doctor", the mayor of Catania and Euro-MP for Forza Italia, Umberto Scapagnini. Whilst expressing doubts about the Moratti reform, Scapagnini told the story of his son, an associate professor in the USA, who asked him: "Papa, what shall I do, return to Italy?," to which *papa* Scapagnini firmly replied "Stay where you are!"[20]

These aspects are interesting because they clearly conflict with an image, hitherto assiduously cultivated in public, of a "depoliticized science" fully able to regulate itself and falling outside the categories typical of political conflict, to which it not infrequently attributes a value inferior to scientific discussion.[21] According to the physicist Carlo Bernardini, the attempts by politicians to manipulate research are "despicable". The endeavour itself to place protest within the frame of "science against politics" is inevitably undermined by the cleavages just described.

Overall, at precisely the moment when scientists denounce "the invasion of science by politics" and take to the streets to assert their right to take research decisions autonomously, they unwittingly reveal the increasing permeability of the boundaries between the two spheres, and that the nature of research problems is wholly "political". Whilst unprecedented in its advancement of claims through militancy, the scientists' exposure within the public debate can be linked with the various movements which in the post-war period, and beginning with physics, highlighted the worries of scientists about the political and social implications of their work (Union of Concerned

[17] L. Salvia, *Corriere della Sera*, 2 February 2003.

[18] L. Salvia, *Corriere della Sera*, 2 February 2003.

[19] C. Di Giorgio and G. Mola, *la Repubblica*, 13 February 2003.

[20] F. Cavallaro, *Corriere della Sera*, 4 February 2003.

[21] On the media's tendency to represent scientific debate using the model of political reporting, for instance, by balancing one scientific position with another in terms of "for" and "against", *see* Dearing (1995); *see also* Chap. 1.

Scientists, Bulletin of Atomic Scientists, or the Pugwash Movement, which was even awarded the Nobel Peace Prize). Indeed, the public understanding of science movement itself is indicative of the public involvement of researchers in support of a specific vision of the relationships among science, the public, and policymaking.

This aspect is connected to the, not accidentally symmetrical, two-way movement between penetration by citizens into research and the increasingly frequent public mobilization of researchers. The presence of the latter in the public arena – specifically with the recent protests, but also with ever greater commitment to, for example, advocating the reasons of research among politicians and public opinion –is part of the complex dynamics that induce citizens to demand greater involvement in technoscience issues. The more researchers mobilize themselves in forms typical of political lobbying, the more they are perceived as litigants, the more the technocratic view that technoscience issues should be delegated to experts loses its legitimacy.

Whence derives a further, apparently paradoxical, generalization which concerns the "open-ended" nature of participation processes. The frequent public exposure of researchers in support of the demands of technocracy (more resources, more listening, and greater delegation to experts) has stimulated demands for involvement by citizens. It has thus contributed to unhinging a relationship between experts and public founded upon that same technocratic model. Vice versa, it is not implausible that pressures to involve citizens in technoscience issues fuel the perception among experts that they must appear in public to respond to the worries of citizens, or to mark out their role and remit.

In general, it is important to understand that participation by non-experts in technoscience and the participation of scientific experts in the public debate are largely two sides of the same coin: they have common origins and they feed each other. At the same time, both phenomena further erode the bases for appeal to the traditional forms of technocratic delegation and of linear interaction between politics and expertise, making the profound decisional impasse on issues to do with technoscience even more evident.

Chapter 4
Beyond Technocracy: Democracy in the Age of Technoscience

It is easy to dodge our responsibilities,
but we cannot dodge the consequences
of dodging our responsibilities.

(Sir Josiah Stamp, speech to the British Association
for the Advancement of Science, 1934)

4.1 Beyond the Illusions of Technocracy

Why is the technocratic response unable to deal with the problems of technoscientific in contemporary societies? Certainly not because of the obtuseness of citizens or the reluctance of political decision makers to heed the opinions of experts. It is very likely that political decision makers would be well pleased to off-load onto a convenient expert – as for that matter they regularly did until a few decades ago – responsibility for deciding whether to authorize a transgenic flour or where to locate a nuclear waste disposal site. The problem is that this is no longer possible because of the factors that I mentioned when discussing the transformations in scientific expertise and its increasingly less monolithic perception among the general public.

It is not possible because the sacred aura of science as a sphere of neutral action *super partes* science has been eroded by the changes which have marked the advent of the post-academic phase and by phenomena such as the increasing public mobilization of researchers. This erosion is ongoing in a multiplicity of contexts, from the environmentalist movements which enlist or dispute expert knowledge to the judicial authorities. And it is also due to the media's increasingly pervasive role in questioning policy decisions, their relationships with expertise, and their influence on the selection of experts in the public arena according to their own criteria and production routines, rather than those of the scientific community.

It is not possible because the increasingly public exposure of researchers and their institutions distinctive of post-academic science has created communicative short circuits between research and the public which often preclude mediation by policymakers.

Finally, it is not possible because non-experts increasingly demand (and are granted) – which is another epochal change – participation, involvement, or "voice"

M. Bucchi, *Beyond Technocracy: Science, Politics and Citizens,*
DOI 10.1007/978-0-387-89522-2_4, © Springer Science+Business Media, LLC 2008

in highly complex technoscientific issues. We have seen several examples, above all since the 1990s: the mobilizations of patients associations and activists which have profoundly influenced research on AIDS and genetic diseases or the Di Bella case which in Italy dramatically highlighted the permeability of the boundaries among science, society, and politics.

One may disapprove of these phenomena as long as one bears in mind that the same conditions which today make the public discussion of biotechnologies possible have also generated funding for research on cancer and AIDS on a scale unprecedented in the history of biomedical research, on the basis of those "social and cultural reasons" which many deplore. These same conditions have led to the active involvement of groups of non-experts, like patients' associations, in definition of the agendas and priorities of numerous sectors of such research. For example, without the pressure of environmentalist movements and of a public opinion sensitized to environmental damage, those scientists who in the mid-1980s theorized a relationship between chlorofluorocarbons and depletion of the ozone layer would not have seen a hypothesis still being discussed by specialists so rapidly endorsed by policymakers (Grundmann and Cavaillè 2000).

4.2 Will Bioethics Save Us?

No less common than the appeal for experts to be assigned a more pervasive role (and, indeed, sometimes complementary to it) is the attempt to respond in ethical terms, particularly in the specific cases of numerous recurrent issues concerning the life sciences, to technoscience crises and conflicts. Although a (bio)ethical response is often common to positions which are very different in content, it is difficult to deny that the word "bioethics" has in recent years become a sort of magic spell, a shield against decisional dilemmas and potentially divisive issues. Setting up a bioethics committee or commission is the typical response – in certain cases, also required by law – made by research institutes, health boards, and government institutions to experimental research and to the testing of drugs on humans and animals, or in general to lines of research and technological innovations with major implications.

Generally speaking, this response basically conflicts with the process whereby a shared moral vision on which to base universally approved constraints has evaporated in the industrialized West. It also accounts for the often feeble impact of the opinions expressed by bioethics committees and commissions. A variant of this solution consists in transferring the ethical dilemma to a particular category of actors: scientists, for example, or the users of a new technology.

In the former case, control is delegated to the codes of behaviour adopted by professional categories. This is frequently done by researchers – among them in Italy the Nobel Prize-winners Dulbecco and Levi Montalcini, who, for instance, say "yes to therapeutic cloning, no to reproductive cloning", or "yes to the research use of embryos which would otherwise be destroyed, no to the creation of embryos ad hoc".[1] In this variant, the ethical solution does not significantly diverge from the technocratic one: the only difference being that the trust is placed not in the knowledge and decision-making competence of experts but in their moral and professional codes.

But this trust in self-regulation by the community of researchers neglects the fact that science today is no longer a "liberal profession" practised by a small number of insiders, nor is it the big science of the post-war period firmly controlled by national governments. Rather, it is a complex enterprise involving large international research groups whose members must answer not only to their own consciences but often, also to the legitimate expectations of shareholders, without which, moreover, numerous important scientific advances would not have been achieved, most notably the mapping of the human genome.

Another obstacle to ethical, self regulation, is that the time horizon of the consequences of discoveries and innovations is by now largely beyond the individual researcher's range and control.

Hans Jonas has argued that traditional ethics, which concerned itself with actions circumscribed in time and space, and considered human action on non-human objects to be irrelevant, is of scant use today, given that science and technology enable action of an extent and impact which was once inconceivable. Whence derives the central importance for Jonas of the concept of "responsibility" (*see* Jonas 1979). Yet it is hard to locate responsibility in post-academic science. The articles in the journals *Nature* and *Science* which, in 2001, presented the results of the project to map the human genome were signed, respectively, by 275 and 250 authors. Who among them could have assumed responsibility for the effects of a research project presumably distributed among many levels and conducted over a long time span?

To what professional ethos can such responsibility be entrusted if the scientific community, as we have seen, is disintegrating in terms of its reference values and the conception of the researcher's role? Not to mention the geographical horizon in a time of globalized research: as events of recent years show, it is not unlikely that research stigmatized by researchers in Europe or the USA, as in the case of cloning for reproductive purposes, will find asylum, so to speak, in North Korea or in Dubai.

The ethical solution risks removing the complex issues of technoscience from the hands of the omnipotent technocrat and consigning them to a tenuous phantasm.

In a second variant of the ethical response, the ethics that matter are those of whoever is empowered to decide whether to use, for instance, an assisted fertilization technique. Such ethics may be those of a specific religion, for example, Catholicism, or the ethics of the single individual, as in the position put forward by the Radical Party in Italy on the occasion of recent debates in that country (*see* Capezzone 2004).

Interestingly, the theme of public information is thus treated in reverse to the technocratic thesis. The technocrats emphasize the ignorance of citizens in order to

[1] "I am in favour of using surplus embryonic stem cells which must otherwise be destroyed. I am against the idea of producing embryos in order to use their stem cells": Rita Levi Montalcini, *Corriere della Sera*, 9 October 2004; *see also* Dulbecco (2004).

argue that they must be educated into accepting the opinions of experts. Those who advocate an ethical solution centred on the recipients of technoscience underestimate the recipients' misinformation in order to maintain that there is no better judge than the individual confronted by specific technoscientific dilemma.

Although it highlights a crucial aspect – that of individual rights and their historical transformation – the ethical stance seemingly neglects the social (and therefore political) relevance of how individual choices are made. This is easily understood when one moves away from the specific case of assisted fertilization to look more generally at the problem of decisions to do with innovation in the medical-scientific field. Let us suppose, as not rarely hypothesized by authoritative scientists, that it will become possible to intervene in the human genome in order to combat ageing and, ultimately, to postpone death (*see*, e.g., Cavadini 2005).[2] This innovation will seem entirely advantageous and legitimate to the single individual, for whom the prolongation of his/her life will not seem prejudicial to the rights of other individuals. Yet it is evident that if this innovation is applied on a large scale, and it seems difficult to discriminate eligibility for it if individual rights are to be upheld, it might have devastating effects for society and the environment. In any case, the decision to delegate particular issues to the individual ethical sphere is a wholly political decision.

The sharp distinction between the two categories – the producers and users of technoscience – breaks down, as we have seen, in face of the new configurations of technoscience in contemporary society. Companies, patients' associations, and environmentalist movements have become increasingly involved in the definition of the technoscience agenda. So in what ethics and responsibility should we place our trust? Those of the shareholders of Celera Genomics, Inc.? Those of the patients and activists pressing for the rapid trialling of gene therapy or for accelerated approval of anti-AIDS drugs? Those of the environmentalists who enlist researchers and produce data and reports to back their arguments?

And to what ethics and responsibility should we remit the multiple implications of genetic research for people's lives, for instance, in regard to the right (legally recognized in some countries) of a person born by heterologous fertilization to undergo – once they have reached majority age – a DNA test so that they can discover the identity of their biological father? Or undergo genetic screening for vulnerability to certain inherited diseases? To the doctors and healthcare personnel that administer such tests? To the single individual who purchases a paternity test kit on the Internet? (*see* Chap. 2). To the minor born with assisted fertilization who can later track down his/her biological father by means of a home DNA test and by browsing among internet databases (Motluk 2005)?

What happens, in fact, is that decisions on these matters are often taken locally, contextually, and pragmatically, and in response more to organizational needs that to explicit criteria of principle. This, for example, is the case of decisions taken in hospitals on whether to turn off life support for patients with no hope of recovery (*see* Anspach 1997; Nyman and Sprung 2000).

---

[2] Haseltine, interviewed by *L'Espresso*, 2 October 2003.

Note that this is not to deny the value of bioethical reflection; nor is it to argue that bioethics committees are useless. The point being made is that the ethical approach and the relative committees have often been used, even unconsciously, to relegate the concerns of citizens to a delimited space, where they are restricted to accessory aspects to do with values and emotions, and in an ambit sharply distinct from that of technical-rational matters, so that their intellectual validity is negated (Wynne 2001). Simply put, what I said earlier about initiatives for the promotion and communication of science holds here as well: they serve many purposes, but meeting the challenges raised by technoscience in democratic societies is not one of them.

4.3 Why Are Citizens Against Biotechnologies?

Decision-making uncertainties in policymaking inevitably interact with the public perception. From this point of view, asking why technocracy and ethics are unable to furnish satisfactory diagnoses of. and therefore solutions to, conflicts on technoscience amounts to asking (to stay with the issues that most closely embody these conflicts) why, after years of information and communication campaigns on the potential benefits of agro-food biotechnologies, the majority of European citizens are still hostile to them. Or why a referendum in Italy rejected the regulation of assisted fertilization techniques and experiments on embryonic stem cells.

We well know the technocrat's answer: citizens are opposed to GMOs (or experiments on embryonic stem cells) because they do not understand their benefits: because they are ignorant and confused and have been frightened by the media. The answer in ethical terms is equally clear-cut: citizens are opposed to experiments on embryonic stem cells because they adhere to a particular moral doctrine, for instance, the precepts of the Catholic Church.

We have already seen that the explanation based on misinformation and preconceived hostility to science does not stand up. Nor does the role of religious belief appear to be decisive.[3] An alternative hypothesis, which takes account of the scenarios described thus far, as well as most recent studies in this specific area, is that such hostility is deeper-rooted and does not concern biotechnologies alone.

This hypothesis relates to the nature of the questions at issue. Technocrats understandably view GMOs only in technical-scientific terms. They thus loudly proclaim: "Let us not concern ourselves with economic, social and cultural aspects; let us concentrate only on what the science says."

Nevertheless, in the light of the changes described in the previous chapters, this appeal is ingenuous as well as futile. First, technoscientific issues configure

[3] For example, only 2% of Italians believe that the Catholic Church should decide on issues concerning biotechnology. Moreover, the majority of the attitudes expressed did not differ significantly between Catholics and non-Catholics, and on some questions (e.g., about medical biotechnologies), Italians were on average more tolerant than the citizens of non-Catholic European countries. *See* Observa (2005) and Eurobarometer (2003).

themselves increasingly less in the public discourse as purely scientific questions, and increasingly more as "hybrid" ones in which scientific data, economic interests, social priorities, moral values, and cultural dimensions interweave. One cannot detach research on AIDS from the disease's social impact, or research on the hole in the ozone layer from environmental worries and mobilizations – just as one cannot separate citizens' opinions on agro-food biotechnologies from their views on the role of the multinationals, the relationships between Europe and the USA and between the industrialized countries and Third World, and on the role of food in their traditions and identities. Suppressing these dimensions of the public debate in an attempt to make the citizen focus on technical aspects is like trying to get a horse with a lame leg to walk better by plastering the other leg.[4]

Likewise, one cannot separate an attribute like trust from cognitive aspects, as the technocratic vision seeks to do. Trust is not a mere surrogate or a "functional substitute" for knowledge, which stands in its stead only when issues exceed the public's competence and it must rely on the experts (Wynne 2001:59). On the contrary, trust is an essential part of every process of knowledge transmission and validation, among experts just as much as among non-experts.[5] The loss of trust entails a loss of cognitive authoritativeness. Many of the citizens who opposed vivisection in Victorian England and had lost their trust in the scientific knowledge obtained with animals' experiments were also favourably disposed towards different kinds of alternative medicine like homeopathy (Mackenzie 1996).

In general, the public's opinion on the acceptability of the risks of a potential technoscientific innovation cannot be separated from its judgements on that innovation's implications and the importance of its purposes. The point is not (contrary to the frequent claim) that the inherent uncertainty of expertise makes furnishing the certainty that the public craves impossible. We are all accustomed to taking decisions in conditions of uncertainty during our everyday lives; all the more so now that uncertainty has become an endemic feature of numerous areas of social life, besides that of technoscience. The point is that "technical" uncertainty must be managed jointly with uncertainty about the implications and purposes of technoscientific innovation. Only if the latter serves convincing purposes will the public be prepared to tolerate a degree of uncertainty in terms of potential risks and unexpected consequences. In other words, the response to such potential risks

[4] This analogy is not intended to attribute some sort of "animalism" to the public debate, although there is a rich tradition of metaphors drawn from the animal kingdom to denote the relationship between rulers and ruled: the people as a "horse to be tamed"; *see* Schiera (1999), esp. Chap. 5; or the media (or public opinion) as a "beast to be fed", *see* Shudson (1995).

[5] Experts themselves frequently resort to fiduciary elements such as institutional membership, inclusion in special networks, or indeed nationality and personal acquaintance in order to gauge the reliability of a researcher's results, especially in the case of controversies which simultaneously involve the results obtained and the methods used to obtain them: *see* Collins (1985, 2004). As some members of the scientific community have pointed out "In practice, only a very small proportion of the experiments and observations reported in the scientific literature are actually replicated by other researchers". (Ziman 2000:99).

cannot be separated from the answers to the questions "Why are we are doing this?" and "Is it worth it?" (Wynne 2001:66).

It is the public's perception of its ability to handle this hybrid character of contemporary technoscience – and not just specific issues like transgenic food or experiments on embryos – which shapes its attitude towards these and similar issues.

Where highly complex issues like biotechnologies are concerned, public opinion finds conventional forms of democratic representation and political decision making to be inadequate, lacking in transparency, and above all unable to cope with a science that has seemingly lost its "independence", impartiality, and inner cohesion. Not coincidentally, it is precisely those people who perceive scientists as disagreeing on GMOs who express most scepticism about agro-food biotechnologies (Bucchi and Neresini 2004a).

Note that in this respect, too, the traditional technocratic explanation can be reversed. Cases like the BSE emergency cannot be interpreted solely as generic "crises of trust" in the relationship between the public and scientists; on the contrary, they fuel the perception that decision making is unable to cope with technoscience, and especially that technocracy is unable efficiently to manage technoscience's uncertainties and implications.

One might even go so far as to claim that citizens do not fear biotechnologies as such. Rather, what they fear that there will be no suitable forums in which to address the multiple facets of technoscience dilemmas, no reliable decision-making procedures able to protect them, and no persuasive and practicable ways in which they can access information about the criteria used to select the experts and counter-conflicts of interest. More than one-fifth of Italians believe that decisions about biotechnologies should not be delegated to scientific experts, rather that "all citizens" must be consulted. Another 5% believe that "the potential recipients of biotechnology applications" must be consulted, and 14% believe that, at the moment, "nobody is able to decide" (Observa-Fondazione Bassetti 2003; Observa 2005). In this case, too, the greater the perception of divisions within the scientific community, the greater the inclination to revoke a mandate giving the experts a free hand, with the preference instead of involving all citizens (Bucchi and Neresini 2004a).

In the absence of such guarantees, and in the absence of a convincing way to delegate technoscience issues to the ethics of researchers, partly because of the reasons already given, partly because researchers themselves are considered to be "interested parties", many citizens prefer to say "no" to the technoscientific innovations proposed.

This position can be defined in the terms of political-pragmatic interpretation of what is known in policymaking as the "precautionary principle": do not permit innovations until their absence of risk has been proven. In this case, the precautionary principle is strictly "political": do not permit innovations if decision-making procedures are not transparent and do not guarantee the due involvement of citizens. To cite Latour, if the slogan that accompanied the transition from absolutism to representative democracy was "no taxation without representation!", today the eclipse of technocracy could be accompanied by the slogan "no innovation without representation!" (Latour 2004).

It is for these reasons that the current crisis extends well beyond the individual issues of GMOs or research on embryos to lay bare the increasingly difficult relationship among scientific expertise, political decision making, and forms of democratic representation.

4.4 Knowledge Is Power

Both the technocratic and ethical responses attempt to redirect tensions arising in political decision making elsewhere, respectively, confining them within the space of expertise and that of ethics. When these attempts fail – as they have done in the case of GMOs in Europe – after years of expert assessment and decisional deadlock, the dilemma is further converted into a mere matter of consumer preferences: it is to the consumer that the regulations on labelling delegate the choice between conventional foods and transgenic ones.

But whilst bioethics can perform a crucial role in showing us the options available and the values involved,[6] only politics can take the decisions, and in doing so balance interests, values, and (today more than ever) scientific expertise, assume responsibility for current generations (and increasingly those of the future) without off-loading it onto one of the parties concerned (the researchers, the patients, or the consumers), as happened in the case of the European regulation on GMOs: "Here's a nice label, now get on with it". And only politics can deal with the multiplicity – perceived at least – of the scientific opinions and moral standards that crowd out the technocratic and ethical options.[7] Finally, only politics can answer the crucial question behind every citizen's attitude towards technoscientific innovation: "Why are we doing this?" (*see* Chap. 3; Bucchi and Neresini 2004b).

The conclusion that the only way out the impasse of decisions on technoscience is a political one naturally arouses numerous misgivings, especially among those – including numerous representatives of the research community – for whom politics have assumed a markedly negative connotation (the "despicable hands" of politicians meddling in research, as a scientist put it at the time of the mobilization against the Moratti bill to reform the Italian research institutions).

Yet these misgivings are justified only if politics is viewed in purely negative-regulatory terms, as a sort of police constable hurrying after the constant progress of science to remedy its harmful effects. This conception of research policies was bred from the first mobilizations and concerns about damage to the environment and health which arose in the industrialized world during the 1960s and 1970s. The role of science and, in particular, its relations with political power and society were

[6] The reference here is to Reich's (1978) classic definition of bioethics.

[8] The paradox of transferring dilemmas of democratic politics to ethics has already been pointed out by Lippmann: "while possibly it may be the aim of political organization to arrive at a common standard of judgment, one of the conditions which engenders and makes political organization necessary is the conflict of standards." (Lippmann 1925:31).

contested by a section of the scientific community itself and by social movements like feminism and environmentalism. The progress of research began to be seen as a source of, rather than as a solution for, social problems which had moved up the public agenda: the proliferation of nuclear weapons, international conflicts and the arms trade, the deterioration of the environment, and gender discrimination. This unprecedented "social pressure" broadened the compass of research policies to include control and regulation. In several countries, in fact, agencies were created to monitor the impact of technology on the environment, and in the USA, the Office of Science and Technology Policy was abolished and replaced with the Office of Technology Assessment instituted by Congress in 1972. Numerous initiatives were launched, especially at the local level, to involve non-scientists in the devising of research policies on matters of particular public importance like nuclear energy or genetic engineering.

As Dorothy Nelkin has observed, whilst the motto of the 1933 Chicago International Exposition held to celebrate "a century of progress" was "Science Finds – Industry Applies – Man Conforms", a World Fair organized in the 1970s would certainly have adopted the slogan "Science Finds – Industry Applies – Man Controls" (Nelkin 1977:393). But a conception of this kind has become just as inadequate today as used to be the conception of politics as the mere means to allocate resources typical of the period of big science. This is not only because, as has been rightly pointed out, the challenges of technoscience shift the attention of politics from the regulation of normality and identity to the regulation of novelty and differences, or because knowledge, having assumed in contemporary societies a role comparable to that of labour and property in the traditional capitalist societies, is bound increasingly to attract the regulatory attention of power (Stehr 2005). Moreover, the processes themselves of technological innovation as embodied in the everyday routines of non-experts transcend the boundaries of traditional control measures: consider how easy it is for a citizen of France – where a 1994 law on bioethics bans any DNA test for purposes other than research or judicial inquiry – to go online and purchase a paternity test kit from an English, Spanish, or Dutch laboratory (*see* Chap. 2).

Contemporary technoscience – as a "generalized capacity to act" (Stehr 2005) – together with proximity to applicative contexts, the incorporation of the attitudes and demands of non-experts, and the active presence of the research community in the public arena, undermines the traditional distinction between knowledge and power: a distinction, moreover, deployed with great rhetorical elasticity in the public discourse: whence philosopher Jerry Ravetz's aphorism "Science takes credit for penicillin, but Society takes the blame for the Bomb." The same distinction is often emphasized in a deceptive opposition whereby "knowledge can develop only outside the injunctions [of power], its demands and its interests" (Foucault 1975, tr. eng. 1995:27). Put in the terms of an antithetical and zero-sum choice between knowledge ("freedom of research"), on the one hand, and power (regulatory constraints) on the other, the crisis of technoscience is bound to be stuck in a blind alley; all the more so when this artificial opposition between knowledge and power is shared by both the technocratic and the ethical responses. The former is unaware that its idealized "knowing subject" is impregnated with power – the power of economic groups, pressure groups, and political decision makers. The latter does not realize that its idealized "deciding subject" (in the "secular" ethical version) or "nature subject" (in the Catholic version) exists in so far as it is already impregnated with a technoscience that pervades every area of everyday life, from consumption to communication to health.[8] And if it is not feasible, for the reasons already given, to *discipline* society into conforming with technoscience, it is today equally impracticable to *discipline* technoscience in the manner proposed by the economist Sir Josiah Stamp in 1934 speech to the British Association for the Advancement of Science. Stamp called for "a moratorium on invention and discovery", in order that man have a breathing spell on invention and discovery in which to adjust his social and economic structure to the constantly changing environment with which he is presented by the "embarrassing fecundity of technology" (cit. in Merton 1938b:262).

The composite and fragmented nature of post-academic science, its suffusion by multiple and conflicting forces and pressures (among them those applied by non-experts), make it hardly compatible with either rigid self-discipline or the mere application of power in the strict sense. Neither the latter nor scientists themselves would today be able to "halt" technoscience even if they wanted to; their laboratories are too susceptible to the expectations of patients, activists, business persons; citizens, consumers, and policymakers are too dependent on technoscience in their activities and lives.

4.5 The Presumed Neutrality of Technoscience

"Suppose an angel came down from heaven and promised the people of the USA a marvellous invention. It would simplify their lives; enable the injured to receive quick treatment; decrease the time of transportation by a large magnitude; brings families and friends closer together [...] in grateful return for this boon to human welfare, the angel demanded that every year 5,000 Americans be put do death on the steps of the Capitol."

It is said that the philosopher Morris Raphael Cohen would begin his courses on ethics with this anecdote. After asking the students how they would answer the angel and getting them to discuss the question, Cohen would then remind them that 5,000 people in the USA die in road accidents every year (Gusfield 1981:3–4).

[8] In Foucalt's terms: "We should admit rather that power produces knowledge (and not simply by encouraging it because it serves power or by applying it because it is useful) that power and knowledge directly imply one another; that there is no power relation without the correlative constitution of a field of knowledge, nor any knowledge that does not presuppose and constitute at the same time power relations. These power/knowledge relations are to be analysed, therefore, not on the basis of a subject of knowledge who is or is not free in relation to the power system, but, on the contrary, the subject who knows, the object to be known and the modalities of knowledge must be regarded as so many effects of these fundamental implications of power-knowledge and their historical transformations." (Foucault 1975, tr. eng. 1995:27–28). On the importance of contemporary technoscience in shaping individual experience through practices such as consumption, *see* Michael (1998) and Elam and Bertilsson (2003); through the practices of health care, *see* Price (1996).

Another obstinate illusion that prevents us from escaping from the technoscience impasse is the conviction that the results of research and technological innovation are morally and politically neutral. What makes them "good" or "bad", it is alleged, is the use made of them by society and politics. This illusion is exemplified by Alfred Nobel, the "repentant" inventor of dynamite who, to counterbalance the perverted uses made of his innovation, provided in his will for a prize which would honour the scientists making discoveries and inventions most beneficial to humankind. Subsequently, and especially in the period between the two world wars, the Nobel Prize continued to foster, sometimes paradoxically, an ideal of neutrality at another level: that of science's indifference to conflicts among nation-states. Especially in the nineteenth century, when it was epitomized by episodes like the honouring of the English chemist Humphry Davy by the French Academy of the Sciences at the height of the Napoleonic Wars, this neutrality helped engender the antithesis drawn between "the nasty partiality of politics and the tranquil rationality of science" (Friedman 2001:81). It was this image of neutrality that science had to maintain through the bloody wars of the following century, and which corresponded to the neutrality of the country, Sweden, in which the Nobel Prize was awarded. Thus, for instance, at the end of both the World War I and II, the Swedish Academy of Sciences had no hesitation in honouring two achievements strictly connected with military enterprise: that of the chemist Fritz Haber, who had headed the German military programme to develop gases for use in chemical warfare (1919), and that of the chemist Otto Hahn, who discovered nuclear fission, and who received the prize only 4 months after the first concrete application of his discovery had destroyed the city of Hiroshima (1945).[9]

The illusion of neutrality fits easily with both its technocratic and ethical versions: according to the former, technoscience is always neutral; according to the latter, there are "good" and "bad" producers and users of technoscience.

The hybrid nature of the objects of contemporary technoscience and the processes of co-production that give rise to it constantly contradict this illusion of neutrality. A safety belt which locks automatically, or a car programmed not to start if the driver has not fastened his/her safety belt, ensures the adoption of certain behaviours deemed safer or socially desirable. They are no longer mere technological objects but hybrids embodying a moral and socio-political vision. Thus, a magnetic key card which turns off the lights of a hotel room or a protection code which prevents the copying of a CD configures the profile of what the well-behaved hotel guest or music consumer should be like (Friedman 2001:81). Likewise, an anti-AIDS drug approved in half the time that the experts and the competent authorities envisage, or the over-riding priority given for many years to research on AIDS to the detriment of diseases characterized by much higher mortality rates, embodies a particular vision of social problems and their relative importance.

The technocratic option, moreover, does not eliminate the political dimension of technoscience, but simply makes it opaque. This brings in Daniel Callahan's

---

[9] On the "cultural politics of neutrality" connected with the Nobel Prize, *see* Friedman (2001). On the award to Hahn, *see also* Bernstein (2001) and Maurizi (2004).

analysis of research and technology in the medical-health field. The last century saw advances in this sector which raised average life expectancy in the industrialized countries from around 45 to 75–80 years. Yet, according to Callahan, the point has now been reached at which the frenetic development of medical knowledge, and above all technology, is not matched by significant improvements in the quality of health, but rather by an unstoppable increase in the expectations of healthcare recipients. Nevertheless, in the USA, since 1960, the per capita cost of medical technology has increased by 791% and the share of national wealth allocated to health care by 269%. What, according to Callahan, does this mean?

It means that nobody will ever say "that's enough" concerning the opportunity to gain a small marginal benefit in health care, but at extremely high cost to society, until state budgets explode. It means that, in the semblance of inertial drag due to the progress of technoscience, the dream of a perfect body and eradication of the concepts themselves of illness, old age, and death have tacitly become political and social goals of absolute priority, with the consequence that resources have been taken away from other healthcare activities, such as prevention, palliative care, and co-operation with developing countries, or from other societal objectives like education or social security (Callahan 1998).

The same considerations apply to the European Union's recurrent declarations of intent to increase investments in research and development to 3% of each member-state's gross domestic product. This is one of the targets set by the so-called "Lisbon Agenda" as necessary to achieve a "European knowledge society", and ultimately to re-launch Europe's economy, and indeed its society. Of course, it is difficult to find anyone who would dispute for the need to increase investments in research and innovation, and in fact, there is no record of a single voice critical of the Lisbon Agenda by any representative of the scientific community or any European politician. This consensus seems inanely demagogic, particularly since nobody feels invested with the political responsibility to pursue this grand objective. The real question to put to the European heads of government and citizens is this: "What are you are prepared to forgo in order to invest in research and development? Are you willing to reduce spending on education? On health? On pensions?"

The political dimension that emerges from the debates on technoscience is indubitably difficult to relate to the traditional categories of "good/bad", "beneficial/harmful", "right/left", and "conservative/progressive." Every technoscientific object is kept in being by a network consisting not only of researchers and their results but also social movements, patients, journalists, business persons, as well as conceptions of the individual, the family, nature, and society. Vice versa, in the absence of such a network, not even the technically most sophisticated and efficient object can come about. Consider the case of the car technologies developed to prevent road accidents involving pedestrians, which is regarded by numerous institutions as a leading priority of road safety policy. From the technical point of view, devices deemed satisfactory by experts have already existed for a number of years. Yet one of the weaknesses of this innovation is the difficulty of convincing subjects for which it would be of no direct benefit – car manufacturers and their customers – to carry the cost, as well as the scant capacity of the issue to coalesce significant public mobilization to demand legislation on the matter (*see* Hamer 2005).

Does it make political sense for medical technoscience to seek, by concentrating mainly on innovating the technologies available to it, to increase that population's life expectancy by 20 or 30 years, even at the cost of huge economic investments which deduct resources from other activities like prevention or from other important political and social objectives like education? In the Europe of 1900, it probably did make sense, but in the USA of the twenty-first century, according to Callahan, it does not. Cases like that of the car mentioned above are among the trump cards that the supporters of the technocratic point of view play to stress that excessive regulatory constraints and the worries of non-experts are hampering the diffusion of innovations. If the car were invented today, it is claimed, and all these restrictions were applied or all these aspects were considered, permission would never have been given for its manufacture. And yet, one may ask provocatively, are we really sure today – amid current concerns for the environment and energy sources, and with the prediction by international development scenarios of not some millions but some billions of potential car owners – that the car is the technological response to the problem of individual mobility that makes most political sense?

From this point of view, the failure of GMOs in Europe is due less to strictly technical factors (their alleged or real danger to human health), or to moral ones (the greed of the multinationals and the honesty of the scientists involved), than to a lack of political vision among those who have failed to consider – as part of the network necessary to support such an innovation – not only farmers but also consumers, the cultural significance of food in numerous countries, and the rootedness of attitudes which rightly or wrongly lump together the USA, multinational companies, the standardization of lifestyles, and the reduction of biological and cultural diversity.

4.6 The Horse that Knew How to Do Sums

"You have probably heard about horses or dogs able to do arithmetic and to count. They are still to be found in some circuses. The trainer makes them enter, and the audience sets them some simple problems. For example, what is four times three or twelve divided by six. In the former case, the horse starts to stamp the answer with its hoof: one, two, three, four, and so on, until it reaches twelve, and then stops; in the latter case, the dog barks twice. The animals are never wrong (or almost never: but that does not matter, even scientists commit errors). How did the trainers realize that their animals could perform simple calculations? Why did they decide to train them? They started from the assumption that animals have minds and are able to reason [...] We know (or we think we know) that those dogs and horses cannot actually count. They react to imperceptible movements by their trainers. The horse begins to stamp out the answer – one, two, three ... when it reaches twelve the trainer, without realizing it, slightly relaxes. The horse reacts exactly to this relaxation [...] Therefore an idea which 'has imposed itself' because of its 'prodigious capacity to produce results' has proved to be wrong." (Feyerabend 1996:52).

With the metaphor of the horse that (apparently) knew how to count, the philosopher of science Paul Feyerabend intended to show that the extraordinary practical success of modern science did not necessarily entail total endorsement of a conception of the world centred on scientific materialism.[10] Without entering into the merits of the specific philosophical value of this argument, it serves to highlight a significant element in understanding the public dimension of contemporary technoscience, and therefore the dilemmas that it raises.

According to the technocratic thesis, technoscience has changed from being a key resource for policy decisions into a self-referential policy culture which assumes that appreciation of its results automatically entails full endorsement of that culture as an end in itself. And it is to this policy culture – perhaps even more than to the specific contents of research and innovation – to its presumption of hegemonizing the debate and the decisions taken on its behalf, to the characterization of its role in normative terms that the public reacts with annoyance when it expresses its opinions or mobilizes on issues like biotechnology (Wynne 2001). It is by emphasizing its capacity to handle all uncertainty on its own (a capacity contradicted in the eyes of the non-expert by episodes like BSE or nuclear accidents), depicting a passive and cognitively deficient public, and relegating it to the margins of decision-making processes as allegedly incompetent, that the technocratic vision brings into being the very same public scepticism against which it contends.

It should be stressed that the technocratic conception, in the sense with which the expression is used here, does not necessarily coincide with research and innovation. Likewise, criticizing the technocratic option is not to blame the community of experts alone for the impasse in which technoscientific finds itself.

The technocratic option has not only fulfilled the need of experts to emphasize their role and to influence policymaking, or simply to protect their prerogatives against the offensive by the public. In numerous specific cases (and also more generally), delegation to technocracy has been a tempting option for political decision makers, and for citizens themselves, who have seen it as an easy way out of conflicts and the assumption of responsibility. The non-experts have fostered, when they have not actively cultivated, the illusion that a technocratic solution can be found for the dilemmas of technoscience. As we have seen, this has had unfortunate long-term consequences for the experts who have assumed such responsibilities, and yet undeniable short-term benefits for experts individually. Biomedicine furnishes numerous examples of how technoscience has been interpreted – also and especially on the basis of the expectations and demands of its users – as a substitute for other actions and institutions in tackling problems such as social inequality, the imbalances in world economic development, the weakening of solidarity, or the crisis of the family, to the point that the members of community of experts have coined the expression "medicine of desire" to denote the extension of biomedical technoscience well beyond the prevention and the treatment of disease to the satisfaction of

[10] This analogy is not intended to attribute some sort of "animal nature" to contemporary technoscience, although there is a rich tradition of metaphors drawn from the animal kingdom to describe features of the scientific enterprise: for example, the lynx, whose sharp-sightedness the founders of the Accademia dei Lincei wished to attribute to themselves. *See* Freedburg (2002).

people's demands for improvements in their appearance and performance, for instance, through cosmetic surgery or performance-enhancing drugs in sport.[11]

4.7 The Crisis of the "Double Delegation"

The challenges of technoscience are disruptive also in so far as they lay bare the fragility of a tradition – the so-called "double delegation" – on which the contemporary democracies were long based: delegation to professional scientists for knowledge about the natural world and delegation to professional politicians for knowledge about the "socio-political collective" (Callon et al. 2001). The whole of modernity, or, according to some, the illusion of modernity, has been based on this allocation of responsibilities: on the one side, facts, pryons, gene sequences, and stem cells; on the other side, the concerns of ecologists, the interests of cattle farmers, international trade agreements, and the protests of patients. The former are allocated to the laboratory, the latter to the parliament.

Some scholars regard the dispute on the existence of a vacuum, in which Robert Boyle defeated his opponent Thomas Hobbes, as emblematic of the historical process by which this "allocation" has come about. For Hobbes, an accord on knowledge which disregarded the political dimension was inconceivable, whereas Boyle saw his vacuum pump much more as a solid guarantee of common accord, an unalterable fact "whatever may happen elsewhere in theory, mataphysics, religion, politics, or logic" (Latour 1991, tr. eng 1993:18; Shapin and Schaffer 1985).

Purged of every socio-political impurity, of every material interest, of every social element, or value judgement able to contaminate it, science acquires in the eyes of its protagonists, and of society as a whole, the aura of neutrality and independence which is perhaps what many commentators hanker after when they invoke the "freedom of research".

This is what Latour calls the "humiliation of politics" by the philosophical tradition encapsulated in Plato's allegory of the Cave. To arrive at the truth, scientists must free themselves from the "tyranny of the social dimension, public life, politics, subjective feelings, and popular agitation" (Latour 1999, tr. eng 2004:10): a tradition without which science would probably not have progressed and become institutionalized, but which today seems to have reached a turning-point.

The hybrid nature of contemporary technoscience dilemmas and of the forums in which post-academic science develops upset this fragile balance between purification and allocation: parliaments (and the courts) find themselves dealing with embryos, measurements of radioactivity, threshold percentages of transgenic organism; laboratories and scientific conferences are attended by patients, shareholders and stockbrokers, and ecologists. Political decision makers and judges ask scientific

[11] The expression "medicine of desire" was introduced by R. Frydman, one of the pioneers of assisted reproduction in France, *see* Pizzini (1992). *See also* Callahan (1998) and Neresini (2001).

experts whether they should ban a transgenic flour or ratify the Kyoto Protocol. Scientists ask politicians to define what is meant by the "substantial equivalence" of genetic traits, or judges to establish what counts as an original scientific discovery. The "voice of the scientific community" is no longer perceived as an oracle "speaking truth to power" (Wildawsky 1979), but as a discordant chorus which often exacerbates, rather than alleviating, the decision-making impasse of politics. Thus blocked is the trade-off whereby the uncertainty of the political debate was long anchored to the solid rock of expertise. With the aggravating circumstance that the humiliation of politics and its demotion to the role of the clearing house of scientific expertise is principally to the detriment of science itself. Given ever greater responsibilities by public opinion, science is an easy scapegoat as soon as an emergency slips out of control, as dramatically exemplified by episodes like the BSE crisis.

4.8 "Etsi Veritas Non Daretur"

Accordingly, the primary concern used to be whether science, and its autonomy in particular, could withstand the pressures of democratic participation. Only recently has the question been put in reverse: whether politics - and democracy in particular - can withstand the massive injection of technoscience which creates apparently new problems for them (*see*, e.g., Rusconi 2004).

But what if the challenge raised by complex decisions and the "technical reproducibility of life" (*see* De Carolis 2004) represents not a problem to be dealt with but an opportunity to rediscover politics and democracy, not only procedurally but in full and frank discussion between visions of the world and humankind?

When the issue of embryo stem cells was debated in Italy, numerous commentators pointed to Great Britain as an example to emulate (*see*, e.g., Boncinelli 2004). Yet in Great Britain, research on embryos had for almost 30 years been a matter of heated political debate within and without parliament. Already in 1978, following the birth of Louise Brown, the first "test-tube baby", Parliament had discussed the Warnock Commission's proposal that researchers be allowed to conduct experiments on human embryos within the first 14 days of the embryo's life. An analysis of the parliamentary records has shown that both parties (those in favour and those against) recounted "stories" prefiguring scenarios of the future which they believed would ensue from one or the other option. Ironically, it was mainly those in favour of the use of embryos who evoked Frankenstein scenarios in order to show that they were baseless (Mulkay 1997; *see also* Turney 2000). Visions of humankind and visions of its future: what lies in store for us and what we want to achieve? Who, if not politicians, could ask themselves such questions?

Indeed, this challenge may be an opportunity to free and give full realization to politics hitherto obliged to interrogate first religion and then science to understand its own purposes.

The considerable price to be paid for this transition is relinquishment of the comforting reassurance furnished by the technocratic delegation. Of course, this does not mean forgoing scientific expertise, without which, for the reasons already given, politics and public debate itself would today be unthinkable. By calling into question the possibility of a politics reliant on technoscience, in which democracy and rationality necessarily coincide, the crisis of the double delegation – to use another expression of Latour – confronts us with a science "freed from the politics of doing away with politics (Latour 1997:232; *see also* Latour 1999, 2003).

In a series of articles widely cited during the Italian debate on assisted reproduction, Italian political scientist Gianenrico Rusconi argued that the issue should be addressed *etsi deus non daretur*, "as if there were no God": that is, on the basis of a secular principle whereby no party to the debate should seek "to impose by authority the truths of its faith or its beliefs" (Rusconi 2002:670).

Hence, a politics that wants fully to settle accounts with the challenges of technoscience, and the real contribution of post-academic science, without shirking its responsibilities must probably address such dilemmas *etsi veritas non daretur*[12]: that is, without expecting expert knowledge to furnish the illusion of anchorage that has been long cultivated (by politics and public opinion more than by the experts themselves, although the latter have not infrequently used it as an argumentative resource in their relations with policymakers and citizens). A politics of this kind would be doubly and wholly "secular" because it would not delegate its responsibilities either to a specific religious (or ethical) vision or to the arbitration of scientific expertise.

In regard to the crucial issues raised by the life sciences, this entails acknowledgement that the stakes are higher than discussion on the status of the embryo and the opposition between science and ethics, between agnostics and Catholics. For the first time in history, humankind has the capacity to modify its own self rather than the environment. A crucial question is whether we want to take up this challenge or to dodge it. The supporters of permissiveness must have the honesty to recognize that they not only advocate individual freedoms but also a (politically legitimate) vision of the active transformation of humankind. On the contrary, it will be necessary to recognize that the "state of nature" is not an initial given, but rather an ideal (equally politically legitimate) whose pursuit requires just as much research and technology.

The wholly political nature of the decisions to be taken is all the more evident if we consider that the current dilemmas express broader transformations extending not only beyond the possibilities offered by contemporary biology but even those by science and technology. By way of example, recently much discussed has been the announcement by a Hong Kong clinic that it can enable parents to choose the sex of their baby on the basis of complex calculations relative to the period in which fertilization will take place.[13] If this claim is borne out, what will the consequences

[12] It is worth repeating that here – and throughout this book – I am not referring to the concept of "truth" in the philosophical sense, nor in the specifically epistemological one, but only with sociological meaning. Thus, "truth" simply connotes results, opinions, or statements that the actors in question (policymakers and public opinion) deem able to orient their decisions unproblematically.

[13] *Hong Kong clinic says it lets parents pick baby's sex through good timing*, in *The Japan Times*, 18 October 2004.

be? And what will the attitude of the public institutions be towards this eventuality? Does the possibility of determining the sex of our children horrify us only because it involves genetic manipulation of the embryo? Or does it do so because it challenges our convictions of what procreation, heredity, chance, and the individual's lot should be? (*see* Galli della Loggia 2004). Whilst we should not overestimate the role of science in resolving the dilemmas that confront us, nor should we overestimate the responsibility of science for raising such dilemmas. The advances of contemporary biology are part of a much broader array of social and cultural changes affecting the concepts of "person" or "family".

Giorgio Israel has written that "[w]hat we should think about life, death and health, about the way in which we manage the relationship between our body and mind, about what our children mean to us" are not questions which "can be settled by the geneticist" (Israel 2004). They are crucial questions for any society wishing to interrogate its identity and its future. And they are dilemmas which are only apparently new. What is new, and perhaps most bewilders us, is the demise of the comforting and long-cultivated illusion that such matters can be kept out of politics and resolved in merely technical-scientific terms. Like it or not, the moment may have come to resume discussion of the kind of world we want to live in.

4.9 Choosing the World We Want

Phenomena like public participation in technoscientific processes, and crises and conflicts on technoscience more generally, signal the growing pressure for such changes to be brought back into democratic politics, where they may be guided by the science and the economy that modernity strove to exclude from democratic politics (*see*, e.g., Beck 1986). Anything but straightforward, however, is defining what politics and what democracy are configured by such phenomena.

The "anaesthetization" of democracy by the massive injection of technical-scientific expertise – and not just that of the natural sciences: consider economic policy decisions – has not been enough to resolve the crucial dilemmas that confront contemporary societies. However, this is not a sufficient reason for believing that these same dilemmas can be simply resolved by injecting democratic procedures into science; even less so when democracy is understood in its more simplistic meaning of "decision by majority vote". Translated into the option of public consultation by referendum, this method has been repeatedly used in recent years to settle controversial technoscience issues. A referendum halted the development of nuclear energy in Italy (1987); a referendum first opened the way for the creation and patenting of living transgenic organisms in Switzerland (1998) and then imposed a five-year moratorium on transgenic plants (2005); a referendum approved government funding for research on embryonic stem cells in California (2004) and has confirmed such funding in Switzerland (2004); and a referendum maintained a regulation on assisted reproduction and experimentation on embryos in Italy which the promoters of the same referendum judged too restrictive (2005).

Yet the referendum method couples involvement of a citizens' base potentially comprising the entire electorate with the oft-cited problem of the misinformation of public opinion on the specific issues to be decided. This aspect cannot be denied – as those who delegate all decisions to the individual ethics of citizens tend to do – but nor should it be absolutized to invoke the technocratic delegation. Rather, it is a difficulty which politics must address, not as a problem specific to the age of technoscience but as a distinctive feature of democracy in the most general sense. Authoritative political scientists regard it as a major obstacle against the use of direct consultation, especially on complex issues such as the dilemmas of technoscience. As political scientist Giovanni Sartori puts it: "even if the information base of the mass public should remain as it is, it will be a tolerable and acceptable weakness as long as we are dealing with *electoral democracy*, that is, as long as public opinion expresses itself by electing [...] But opinion is not sufficient for the purposes of a referendum-based democracy; knowledge is also necessary. The leap is indeed one of quality; and a huge one at that; and all our knowledge gainsays its feasibility" (Sartori 1993:75, 87). The only additional point to be made concerns, as repeatedly said, the real or presumed novelty or the greater complexity of the contemporary technoscientific dilemmas, compared with those that used to afflict democracy. Was it really so simple for Italian citizens to choose between the proportional and majoritarian electoral systems, as they did with a referendum in 1993, when authoritative politicians and political scientists were (and are) still strongly discussing the implications of one or other system for the stability of governments? Was it so simple for British voters in 1975 to weigh the costs and benefits of their country's membership of the European Common Market?

Besides the intrinsic complexity of such decisions, it is likely that the context in which these decisions are taken has changed, not only because of the already-described transformations in science and in its perception, but also because of the broader changes ongoing in the contemporary democracies. In other words, complex decisions, also those concerning technoscientific matters, used to be treated, on the basis of a certain conception of expert knowledge and its role, *as if* they were simple. *Etsi veritas daretur*: as if the opinion of the expert could unequivocally and unproblematically direct choices.

But the fact remains that treating democracy as "given" with which technoscience must conform springs from an ingenuousness which mirrors that of the technocratic or ethical positions and which want to do the same thing with society and democratic participation. If science changes when citizens enter the research laboratories, the same will happen with equal inevitability to democracy when scientists protest in public or expressly incorporate social concerns into new areas of research such as "ethical" embryonic stem cells (Meissner et al. 2005; *see also* Testa 2006). It is moreover indubitable that science has made a crucial contribution over the last two centuries to democratic politics, not only in terms of specific competences but also (according to some) by shaping the argumentative styles of democracy.[14]

[14] This is the thesis of, for example, Ezrahi (1990), namely that the birth of democracy was due to the "attestive" style typical of the experimental method, as opposed to the "celebratory style" typical of absolutist regimes (Ezrahi 1990, particularly pp. 73 and ff.).

Elements from technoscience increasingly corroborate and support processes of "tacit democratization", which are perhaps less striking but no less pervasive and which redefine the relationships between producers and users in areas such as information technology or the entertainment industry. The case of open source softwares is by now paradigmatic, but consider the extent to which innovations in multimedia technologies have increased the ability of consumers to produce and circulate their own audio or video products, as well as to intervene in the creations of others. The same pressures for greater participation by citizens in technoscientific processes can be viewed as part of a more general critique against the capacity of the traditional democratic forums to include and represent the opinions of citizens in regard to global challenges, where the crucial decisions are increasingly taken at levels not directly subject to the influence of citizens: a "democratic deficit" by now amply thematized and discussed, for instance, with reference to the European and international institutions.[15] Of course, science and technological innovation are not extraneous to these challenges, in that they emphasize and foster processes of inclusion and exclusion which redefine the concept itself of citizenship and the conditions of democratic participation (Jasanoff 2004a). This is an aspect also contemplated by the new European Constitution, albeit primarily in conventional regulatory terms: "It is necessary to strengthen the protection of fundamental rights in the light of changes in society, social progress, and scientific and technological developments" [...]".[16]

In an even more profound sense, the identity of the citizen is increasingly imbued with technoscience elements mediated by practices such as consumption (Michael 1998, 2002). From this point of view, the difficulty of locating an individual responsibility which weakens an option like (bio)ethical one is not specific to technoscience but reflects more general processes of disaggregation and re-aggregation, in their turn fuelled by technoscience, in respect to notions so crucial for politics and the law as the individual or the person. The example of computer technologies is once again emblematic: where can the responsibility (in this case also criminal) for a peer-to-peer file exchange network be located if there is no central server? How can the "multiple or fragmented identities" typical of online behaviour be defined politically and legally (Turkle 1995; Paccagnella 2000)?

In general, the contemporary dilemmas of technoscience call into question the delegation to the traditional democratic institutions no less than they do the delegation to technocracy. This has prompted scholars like Callon to call for a switch from a "democracy of delegation" to a "technical democracy, or more exactly to render our democracies able to absorb the debates and the controversies provoked by the rapid advances of science and technology" (Callon et al. 2001:23–24). The simultaneous fragility of both sides of the "double delegation" has been made especially apparent by cases like Iceland's *Health Sector Database*, when that country's

[15] For an overview of this problem with reference to the European institutions, *see*, for example, Burns and Andersen (1996).

[16] *Treaty Establishing a European Constitution*, Brussels, 13 October 2004, 71.

parliament decided, by a tiny majority, to grant a private company (*deCode Genetics*) exclusive rights for 12 years to construct, maintain, and sell access to a database containing all clinical, genealogical, and genetic data on the Icelandic population; a decision which provoked great controversy (Arnason and Arnason 2001).

The dilemmas of technoscience do not only arise because research and innovation have accelerated their pace and increasingly raise awkward issues. They also spring from the need of contemporary democracies to tackle changes in technoscience in a profoundly different democratic context.

This is demonstrated by the fact that where there is no democracy, or more concretely, in the absence of some of the elements (like the media or public participation) crucial for both transformative processes, these dilemmas do not arise and the technocratic circuit still operates with all its force. Examples are North Korea, Cuba, and partly China, where issues like GMOs or experiments in reproductive cloning have not aroused the slightest controversy (*see* Lévy-Leblond 2003).

In general, the challenges of technoscience undoubtedly provide the opportunity to renew the concept of democracy in a fuller sense, for instance, by defining it as a context of discussion and deliberation within which any discourse or language, not even scientific language, is hegemonic. Thus taken up is Rusconi's recommendation to devise a democracy which is secular not only because it is "non-confessional" but also because it does not rely on the presumed certainties of scientific expertise, which naturally does not mean indifference to its reasons.[17]

Specifically, the challenge consists in closing the rift caused by "double delegation". This can be done by working not on the distinct sides of technoscience and democracy but on the intersection between them. As Sheila Jasanoff has written: "Innovation in natural knowledge and in its technological applications demands a corresponding capacity for social innovation" (Jasanoff 2004a:91). Hybrid problems and increasingly hybrid forms of co-production of technoscience require hybrid decision-making forums able to integrate the traditional compartments (laboratories, parliaments) within which we could - or deluded ourselves that we could - segregate uncertainty about the natural the social world.

It is in this light that we may read Latour's provocative proposal to create "parliaments of things", in which both nature and society can be thoroughly explored, whilst at the same time, new "rules of method" are established for experiments which are no longer solely scientific but socio-scientific as well – like global warming, BSE, or transgenic foods; experiments which break down the distinctions between experts and non-experts, as well as the boundaries of laboratories, which now extend to encompass homes, firms, farms, and hospitals (Latour 2004).

Attempts have been made (more or less deliberately) to meet this demand by creating various forms of participation and democratic decision making on scientific

---

[17] Rusconi's demand that such secular debate should not be "conditioned by the enterprise-science" (Rusconi 2002:671) is therefore perhaps excessive. As said, the public debate is by now imbued with scientific and technological elements. If it were not, there would be no discussion of the dilemmas and crises of technoscience.

and technological issues. Institutionalized forms of public involvement like consensus conferences and citizen panels have long been used in numerous countries, and recently in Italy as well (*see*Chap. 3; Pellegrini 2004). The effectiveness of these initiatives, as we have seen, varies greatly according to the initial expectations, the specific issues addressed, their ability to integrate with traditional decision-making institutions, and the extent to which they meaningfully involve citizens.

Of course, seeking to resolve the impasses of technoscience merely by seating citizens, stakeholders and scientific experts around a table would likely replace the fetishes of technocracy with the fetish, no less utopian and prescriptive, that discussion can lead to rational agreement between the parties.

The foregoing discussion helps highlight a series of misunderstandings that frequently condition the results and the effectiveness of initiatives to promote public participation in scientific debate.

The first misunderstanding is that such forms of institutionalized involvement and participatory decision making must necessarily smooth out all differences among the various subjects involved. This conviction betrays a highly reductive conception of democratic debate, which is instead made up of conflicts, different views on problems, and the solutions to them, and decisions which very rarely satisfy everyone, but which in principle enable the discontented to accept, if not the result, at least the process that has produced it.

The second misunderstanding is that it is possible to respond to the challenges of technoscience, and particularly to demands for public participation, simply by increasing the number of subjects involved in the decision-making process, as if by merely getting them on board, the train of research and innovation can run more swiftly than before. This idea ominously resembles a demagogic version of the missionary vocation of technocracy, where dialogue is replaced by one-way communication. Its likely outcome is the referendum, which, from the strictly quantitative point of view, is the method that produces the greatest public involvement.

Just as access to information should not be confused with the obligation to undergo communication on technoscience, so it makes no sense to confuse the prerogatives of participation typical of a democratic society – and increasingly expressed in technoscience – with the obligation on every citizen to be constantly involved whatever the cost.

What appears on the surface from the crises of technoscience and from the emergence of participatory processes in the same area is not a demand for general and direct involvement, but rather a demand for transparent and accountable decision-making procedures which citizens can access in order to state their own points of view in particular circumstances and under certain conditions. Such decision-making procedures should be able to manage the new conflicts and the new uncertainties without seeking – as technocracy and ethics instead do – to flatten them under the steamrollers of technical competence and individual conscience.

The way out of the impasse of technoscience is not to sit citizens and scientists around a table and organize a cursory chat between them in the belief that all possible perplexity or dissent can be thus dispelled and "business as usual" be then resumed.

Public participation is not a pledge to pay for the progress of technoscience; it is not yet another technocratic fix to straighten the course of a technoscience diverted by social conflicts; it is not a simple button to press, a process which can be activated or deactivated according to the needs of a research policy decided elsewhere. One reason why this does not make sense is that, as said, the participation of citizens in science is already happening, like it or not, and whether or not it is encouraged. The point of view of citizens is not a problem to be worried about in retrospect: society does not get involved when technoscience has already been manufactured. Dialogue with non-experts is not a "social lubricant" for use only when the decision-making mechanisms of technoscience grind to a halt on a particularly controversial issue. Rather, it acquires sense for the various subjects involved only if it works smoothly "in times of peace", and not solely for marginal issues but also, for instance, for costs-benefits analysis (*see* Bijker 2004).

This is not to embellish already concluded technoscientific processes or to give them the licence of "social sustainability". The real challenge consists in bringing, or recognizing when they have already begun, public participation and open democratic discussion into the initial phase of defining the technoscience agenda.[18]

Continuing to consider public debate and participation as being like inviting a guest at the last minute to sample a pre-established menu (however rich and sophisticated) and giving him only the options of saying "yes" or "no" to each dish, can only fuel a vision of politics as the mere "ex post" regulation of technoscience, and with it the likelihood that non-experts will perceive rejecting the products of research and innovation out of hand as the only way in which they can make their voices heard. Thus exacerbated is the mutual distrust that, in the best tradition of the self-fulfilling prophecy, risks contributing to the "clash of civilizations" which rightly worries not only researchers but anyone who understands the importance of research and innovation being fully legitimated and inclusive – both in their contents and processes - on the social and political levels.

The categories of, and oppositions between, "experts" and "non-experts" and "science" and "society" are now otiose in face of a technoscience which, especially in some sectors, is increasingly produced within heterogeneous and unstable epistemic networks of experts, citizens, patient, companies, and activists (Irwin and Michael 2003).

The current transition appears even more radical than the one symbolized, during the last century, by the change of slogans cited by Nelkin (1977). After the slogan of the 1930s, "science discovers – industry applies – man conforms", and that of the 1970s "science discovers – industry applies – man controls", the current slogan could be "science, industry and society discover, apply and decide".

Moreover, hybrid forums and the co-production of knowledge oblige us to recognize that processes of "politicization" in the broad sense are already abundantly in progress. Do they not engage, legitimately, in politics those American scientists who propose a compromise to the Government: "prohibit the use of federal funds

[18] The Anglo-American literature refers to this process as "upstream public engagement".

for deriving new embryonic stem-cell lines but permit these funds to be used to study embryonic stem-cell lines, whenever they were derived, provided the cells were obtained in an ethically acceptable fashion." (Shaywitz and Mellon 2004). And do they not engage in politics those people who want the great emotion aroused by the death of the actor Christopher Reeves – a well-known campaigner for research on stem cells – to convince public opinion of the importance of therapeutic cloning? (*see* Farkas 2004). If physics lost its "innocence" to Hiroshima, biology lost its own in 1974, when, with a historic public appeal, some of its most visible exponents asked their colleagues to interrupt research on recombinant DNA so that its implications could be evaluated.[19]

Should we not therefore recognize that today every technology embodies a vision of the world, and that in every research program, there is already a political programme – that is, a vision of humanity and its place in nature - and on this basis, discuss and take the necessary decisions? Do we want to improve Africa's water resources or to produce people who have no need of water? Do we want more effective disease prevention or people with livers and lungs modified so that they can drink and smoke as much as they want? Discussing, as is done today, GMOs or stem cells, after decades of costly research and laborious experiments, will perhaps serve to cultivate the rhetoric of a "free" science, but it risks being a frustrating exercise, not only for researchers. Bringing democracy into the heart of technoscience – and technoscience into that of the democracy – would force us to ask what vision of the world we can and want to choose and seek to realize. If science has become too important to be left to scientists alone, democracy certainly cannot do without their contribution. The alternative to a technocratic vision is certainly not non-research and non-innovation, but rather a democracy in which technoscience is a valuable resource and not an alibi for evading difficult political choices; a democracy endowed with decision-making forums and processes in which political uncertainty is openly addressed jointly with technical-scientific uncertainty.

But this requires us to leave behind the vision of a zero-sum relationship between technoscience on the one hand and society and politics on the other, where every concession to democratic participation is perceived as a diminution and a contamination of the role of expert knowledge. It is necessary to understand that public protests against GMOs or nuclear power plants are the other side of the process whereby citizens engage in obtaining resources, data, and opportunities for networking on muscular dystrophy research. It is necessary to understand, as it has been rightly observed, that democracy and expertise are antithetical only as long as we restrict ourselves to a vision of democracy as "majority voting" and to a vision of expertise "as a self-referential system in which only peers can recognize and judge each other (Liberatore and Funtowicz 2003:147).

A knowledge society today is *not only* "compatible" with a democratic society. A knowledge society today cannot exist *without* a society democratic in all its mechanisms, including the governance of knowledge.

---

[19] The so called 'Berg Letter' - *see* Berg et al.(1974).

Bibliography

Anspach, R. (1997). *Deciding Who Lives. Fateful Choices in the Intensive-Care Nursery*, Berkeley, CA: University of California Press.

Arena, G. (1995). Trasparenza amministrativa. In *Enciclopedia Giuridica*, IV, vol. aggiornamento, Roma.

Árnason, V. E., Árnason, G. (2001). Community consent, democracy and public dialogue: The case of the Icelandic health sector database. *Politeia*, XVII, 63, 105–116.

Australian Science Indicators Report. (1991). *Section Science and technology news in the media*, Canberra: Department of Education, Science and Training.

Bader, R. G. (1990). How science news influence newspaper science coverage: A case study. *Journalism Quarterly*, 67, 88–96.

Barnes, B. (1985). *About Science*. Oxford: Blackwell.

Bauer, M. (1998). The medicalization of science news – From the 'rocket-scalpel' to the 'gene-meteorite' complex'. *Social Science Information*, 37, 731–751.

Bauer, M., Bucchi, M. (2007). (Eds.) *Journalism, Science and Society: Science Communication Between News and Public Relations*. London: Routledge.

Bauer, M., Gaskell, G. (2002). (Eds.) *Biotechnology. The Making of a Global Controversy*. Cambridge: Cambridge University Press.

Bauer, M., Gregory, J. (2007). From journalism to corporate communication in post-war Britain. In: Bauer, M., Bucchi, M. (Eds.) *Journalism, Science and Society: Science Communication Between News and Public Relations* (pp. 33–52). London: Routledge.

Bauer, M., Petkova, K. (2005). Long-term trends in the public representation of science across the 'iron curtain': 1946–95. *Social Studies of Science*, 34, 1–33.

Beck, U. (1986). *Risikogesellschaft. Auf dem Weg in eine andere Moderne*. Frankfurt: Suhrkamp.

Bellone, E. (2005). *La scienza negata. Il caso italiano*. Torino: Codice.

Berg, P.D. et al. (1974). Potential Biohazards of Recombinant DNA Molecules (letter), *Science* 185 (148): 303.

Bernstein, J. (2001). *Hitler's Uranium Club. The Secret Recordings of Farm Hal*. New York, NY: Springer.

Biezunski, M. (1985). Popularization and scientific controversy. In: Shinn, T. E., Whitley, R.(Eds.), *Expository Science* (183–193). Dordrecht: Reidel.

Bijker, W. (1995). *Of Bycicles, Bakelites and Bulbs*. Cambridge, MA: MIT.

Bijker, W. (2004). *Technological Responsibility and Social Dialogue*, Paper Given at the Conference "La responsabilità tecnologica", Rome, 2–3 December.

Bobbio, L. (2002). Le arene deliberative. *Rivista italiana di politiche pubbliche*, 3, 5–29.

Boncinelli, E. (2004). I precetti e il dolore. *Il Corriere della Sera*, 12 August, 1.

Borgna, P. (2001). *Immagine pubbliche della scienza. Gli Italiani e la ricerca scientifica e tecnologica*. Torino: Edizioni di Comunità.

Bourret, P. (2005). BRCA patients and clinical collectives: New configurations of action in cancer genetics practices. *Social Studies of Science*, 35(1), 41–68.

Boyle, J. A. (1996). *Shamans, Software, and Spleens: Law and the Construction of the Information Society*. Cambridge, MA: Harvard University Press.

Brannigan, A. G. (1981). *The Social Basis of Scientific Discoveries*. Cambridge: Cambridge University Press.

Brint, S. (2004). *Creating the Future: Structures, Sources, and Prospects for the New University Forms*, Research Paper, University of California Berkeley.

Brown, P., Mikkelsen, E. (1990). *No Safe Place: Toxic Waste, Leukemia, and Community Action*. Berkeley, CA: University of California Press.

Bucchi, M. (1998a). *Science and the Media. Alternative Routes in Scientific Communication*. London: Routledge.

Bucchi, M. (1998b). La provetta trasparente: a proposito del caso Di Bella. *Il Mulino*, 1988(1), 90–99.

Bucchi, M. (1999). *Vino, alghe e mucche pazze: la rappresentazione televisiva delle situazioni di rischio*. Roma: Eri/Rai.

Bucchi, M. (2001). Ricerca, politica della. In: *Enciclopedia delle Scienze Sociali*, vol. IX (245–253). Roma: Istituto della Enciclopedia Italiana.

Bucchi, M. (2002). *Scienza e società. Un'introduzione alla sociologia della scienza*. Bologna: Il Mulino.

Bucchi, M. (2003a). Public Understanding of Science. In: *Storia della Scienza*, vol. IX, *La grande Scienza* (811–817) (245–258). Roma: Istituto dell'Enciclopedia Italiana.

Bucchi, M. (2003b). La democrazia alla prova della scienza. *Il Mulino*, 2003(6), 1050–1057.

Bucchi, M. (2004). Can genetics help us rethink communication? Public communication of science as a 'double helix'. *New Genetics and Society*, 23(3), 269–283.

Bucchi, M., Mazzolini, R. G. (2003). Big news, little science: science coverage in the Italian Daily Press, 1946–1997. *Public Understanding of Science*, 12, 7–24.

Bucchi, M., Neresini, F. (2002). Biotech remains unloved by the more informed. *Nature*, 416, 261.

Bucchi, M., Neresini, F. (2004a). Why are people hostile to biotechnologies? *Science*, 304, 1749.

Bucchi, M., Neresini, F. (2004b). Scienza contro politica o politicizzazione della scienza? La riforma degli enti e il dibattito sui problemi della ricerca. In: Fabbrini, S., Della Sala, V. (Eds.) *Politica in Italia. I fatti dell'anno e le interpretazioni* (167–187), Bologna: Il Mulino.

Bucchi, M., Neresini, F. (2006). (Eds.) *Cellule e cittadini. Biotecnologie nello spazio pubblico*. Milano: Sironi.

Bucchi, M., Neresini, F. (2008). Science and public participation. In: Hackett, E. et al. (Eds.) *The Handbook of Science and Technology Studies*, New Edition. Cambridge, MA: MIT.

Burnham, J. (1987). *How Superstition Won and Science Lost: Popularizing Science and Health in the US*. New Brunswick, NJ: Rutgers University Press.

Burns, T. R., Andersen, S. (1996). The European Union and the erosion of parliamentary democracy: A study of post-parliamentary governance In: Andersen, S., Eliassen, K. A. (Eds.) *European Union - How Democratic is It?* (371–397). London: Sage.

Butler, D. (1999). The Writing is on Web for Science Journals in Print. *Nature*, 397, 195–200.

Callahan, D. (1998). *False Hopes*. New York, NY: Simon & Schuster.

Callon, M. (1999). The role of lay people in the production and dissemination of scientific knowledge. *Science, Technology, and Society*, 4(1), 81–94.

Callon, M. et al. (2001). *Agir dans un monde incertain. Essai sur la démocratie technique*. Paris: Editions de Seuil.

Capezzone, D. (2004). No alla morale di Stato sulla fecondazione. *Il Corriere della Sera*, 9 ottobre.

Carrubba, S. (2003). Contro le lobby anti-innovazione. *Il Sole 24 Ore*, 18 maggio.

Castells, M. (1997). *The Power of Identity*. Oxford: Blackwell.

Cavadini, F. (2005). 'Vivremo fino a 120 anni'. I nuovi vecchi saranno così. *Il Corriere della Sera*, 22 ottobre 2005, 29.

Check, E., Dennis, E. (2005). 'Ethical' routes to stem cells highlight political divide. Split opens over methods to create nonviable embryos. *Nature*, 16 October, Resource Document, www.nature.com

Chemin, A. (2005). La guerre de l'Adn. *Le Monde*, 21.

Chia, A. (1998). Seeing and believing. The variety of scientists' responses to contrary data. *Science Communication*, 19(4), 366–391.

Collins, H. M. (1985). *Changing Order. Replication and Induction in Scientific Practice*. London: Sage.

Collins, H. M. (2004). *Gravity's Shadow: The Search For Gravitational Waves*. Chicago, IL: University of Chicago Press.

Collins, H. M. (2005). Heroes or villains? (Review of Duncan, D. E., The Geneticist who Played Hoops with my DNA). *New Scientist*, 20 August, 50.

Comin, D. (2004). R&D: A small contribution to productivity growth. *Journal of Economic Growth*, 9(4), 391–421.

Crane, D. (1972). *Invisible Colleges: Diffusion of Knowledge in Scientific Communities*. Chicago, IL: University of Chicago Press.

Danchin, A. (2000). La storia del genoma umano. *Internazionale*, 341, 30 June 2000, 20–24.

Dearing, J. W. (1995). Newspaper coverage of maverick science: Creating controversy through balancing. *Public Understanding of Science*, 4, 341–361.

De Carolis, M. (2004). *La vita nell'epoca della sua riproducibilità tecnica*. Torino: Boringhieri.

DeFleur, M. L., Ball-Rokeach, S. J. (1989). *Theories of Mass Communications*. New York, NY: Longman.

Della Porta, D., Turrow, S. (2004). (Eds.) *Transnational Protest and Global Activism*. New York, NY: Roman and Littlefield.

Della Porta, D., Kriesi, H., Rucht, D. (1997). (Eds.) *Social Movements in a Globalizing World*. New York, NY: Macmillan.

Dell'Oste, C. (2005). Perché non serve fare i ricercatori duri e puri. *Il Sole24Ore*, Research special, 23 September, 9.

Diani, M. (1995). *Green Networks: A Structural Analysis of the Italian Environmental Movement*. Edinburgh: Edinbugh University Press.

Di Maggio, P., Powell, W. (1983). The iron cage revisited: Institutional isomorphism and collective rationality in organizational fields. *American Sociological Review*, 48, 147–160.

Dini Valentini, T. (1992). *Analisi e comunicazione del rischio tecnologico*, Napoli: Liguori.

Dulbecco, R. (2004). *Scienza e società oggi. La tentazione della paura*. Milano: Bompiani.

Durant, J. et al. (1989). The Public Understanding of Science. *Nature*, 340, 11–14.

Durant, J. et al. (1991). *Europeans, Science and Technology*, Paper Presented to the Annual AAAS Meeting, Washington, DC.

Eberle, C. (1997). *Scienziati in Immagine*. Unpublished MA Thesis, Facoltà di Sociologia, Università di Trento.

Eglash, R. et al. (2004). (Eds.) *Appropriating Technology. Vernacular Science and Social Power*. Minneapolis, MN: Minnesota University Press.

Einsiedel, E. F. (1992). Framing science and technology in the Canadian press. *Public Understanding of Science*, 1, 89–103.

Elam, M., Bertilsson, M. (2003). Consuming, engaging and confrontino science. The emerging dimensions of scientific citizenship. *European Journal of Social Theory*, 6(2), 233–251.

Eltzinga, A., Jamison, A. (1995). Changing policy agendas in science and technology, In: Jasanoff, S. et al. (Eds.) *Handbook of Science and Technology Studies*, Thousand Oaks, CA: Sage.

Epstein, S. (1995). The construction of lay expertise: AIDS activism and the forging of credibility in the reform of clinical trials. *Science, Technology and Human Values*, 20(4), 408–437.

Epstein, S. (1996). *Impure Science: AIDS, Activism and the Politics of Knowledge*. Berkeley, CA: University of California Press.

Espallargues, M. et al. (2005). *Prova pilot del sistema de priorització de pacients en llista d'espera per a cirurgia de cataracta, artroplàstia de maluc i artroplàstia de genoll*. Report for the Health Agency of Catalunya, www.aatrm.net.
Etzkowitz, H. (1990). The second academic revolution: The role of the Research University in economic development. In Cozzens, S. E. et al. (Eds.) *The Research System in Transition*. Boston, MA: Kluwer.
Etzkowitz, H., Webster, A. (1995). Science as intellectual property. In Jasanoff, S. et al. (Eds.) *Handbook of Science and Technology Studies* (480–505). Thousand Oaks, CA: Sage.
Eudes, Y. (2002). Le pirates du génome. *Le Monde*, 18 settembre.
Eurobarometer. (2001). *Europeans, Science and Technology*. Eurobarometer 55.2.
Eurobarometer. (2003). *Europeans and Biotechnology*. Eurobarometer 58.0.
Eurobarometer. (2005). *Europeans, Science and Technology*. Eurobarometer 63.1.
European Commission. (2002). *Benchmarking the Promotion of RTD Culture and Public Understanding of Science*. Luxembourg: Office for Official Publications of the European Communities.
Ezrahi, Y. (1990). *The Descent of Icarus: Science and the Transformation of Contemporary Democracy*. Cambridge, MA: Harvard University Press.
Farkas, A. (2004). Addio a Superman, l'eroe americano che sfidò la paralisi. *Il Corriere della Sera*, 12 ottobre.
Faulkner, W. (1994). Conceptualizing knowledge used in innovation: A second look at the science-technology distinction and industrial innovation. *Science Technology and Human Values*, 19(4), 425–458.
Feyerabend, P. K. (1996a). *Ambiguità e armonia*. Roma: Laterza.
Fleck, L. (1935). *Entstehung und Entwicklung einer wissenschaftliche Tatsache*. Frankfurt a. M.: Suhrkamp.
Foucault, M. (1975). *Surveiller et punir: Naissance de la prison*. Paris: Gallimard. Translated in English. (1995). *Discipline and Punish. The Birth of the Prison*. New York, NY: Vintage.
Fox, D. (2005). Are IVF embryos starved of a vital ingredient? *New Scientist*, 19 February, 7.
Freedberg, D. (2002). *The Eye of the Lynx: Galileo, His Friends, and the Beginnings of Modern Natural History*. Chicago, IL: University of Chicago Press.
Friedman, R. M. (2001). *The Politics of Excellence. Behind the Nobel Prize in Science*. New York, NY: Times Books.
Friedman, S. M. et al. (1986). (Eds.) *Scientists and Journalists. Reporting Science as News*. New York, NY: The Free Press.
Funtowicz, S., Ravetz, J. (1993). Science for the post-normal age. *Futures*, 25(7), 739–755.
Galli della Loggia, E. (2004). La chiesa, i valori, il laicismo. *Il Corriere della Sera*, 6 ottobre.
Gaskell, G., Bauer, M. (2001) (Eds.) *Biotechnology: The Years of Controversy 1996–2000*. London: Science Museum.
Gaskell, G. et al. (2000). Biotechnology in the European Public. *Nature Biotechnology*, 18(9), 935–938.
Gibbons, M. et al. (1994). *The New Production of Knowledge: Dynamics of Science and Research in Contemporary Societies*. London: Sage.
Gleick, J. (1992). *Genius. The Life and Science of Richard Feynman*. New York, NY: Knopf.
Gleick, J. (1999). *Faster. The Acceleration of Just About Everything*. New York, NY: Pantheon.
Goepfert, W. (2007). The strength of PR and the weakness of science journalism. In Bauer, M., Bucchi, M. (Eds.), *Between Journalism and PR: The Changing Scenarios of Science Communication* (215–226). London: Routledge,
Goldberg, D. A. M. (2004). The scratch is hip-hop: Appropriating the phonographic medium. In: Eglash, R. et al. (Eds.) *Appropriating Technology. Vernacular Science and Social Power*. Minneapolis, MN: Minnesota University Press.
Gregory, J., Miller, S. (1998). *Science in Public. Communication, Culture, and Credibility*. London: Plenum.
Grmek, M. D. (1989). *Histoire du SIDA*. Paris: Payot.
Grundmann, R., Cavaillé, J. P. (2000). Simplicity in science and its publics. *Science as Culture*, 9(3), 353–389.

Gruppo, L. (2005). *Il sapere liberato. Il movimento dell'open source e la ricerca scientifica.* Milano: Feltrinelli.

Gunter, B. et al. (1999). The media and public understanding of biotechnology. *Science Communication*, 20, 373–394.

Gusfield, J. R. (1981). *The Culture of Public Problems: Drinking-Driving and the Symbolic Order.* Chicago, IL: The University of Chicago Press.

Hacking, I. (1986). Culpable ignorance of interference effects. In: MacLean, D. (Eds.) *Values at Risk.* Totowa, NJ: Rowman & Allanheld.

Hagstrom, W. O. (1982). Gift giving as an organizing principle in science. In Barnes, B., Edge, D. (Eds.) *Science in Context* (21–34). Milton Keynes: Open University Press.

Hamer, M. (2005). How to stop the slaughter of innocent. *New Scientist*, 27 August, 22–23.

Hansen, A. (1994). Journalistic practices and science reporting in the British press. *Public Understanding of Science*, 3, 111–134.

Haraway, D. (1997). *Modest_Witness@Second Millennium. FemaleMan©Meets Oncomouse™.* New York, NY: Routledge.

Hogan, J. (2003). Experts can't agree on internal radiation risk. *New Scientist*,19 July.

Home, R. (1993). Learning from buildings: Laboratory design and the nature of physics. In: Mazzolini, R. G. (Ed.) *Non-Verbal Communication in Science Prior to 1900.* Firenze: Olschki.

Hooper, R. (2005a). Television Scrambles the Evidence. *New Scientist*, 10 September, 12.

Hooper, R. (2005b). A glimpse of your future for $1000. *New Scientist*, 20 August, 15.

House of Lords. (2000). *Science and Society.* Report of the Select Committee on Science and Technology, Resource Document, http://www.parliament.the-stationery-office.co.uk/pa/ld199900/ldselect/ldsctech/38/3801.htm.

Irwin, A. (1995). *Citizen Science. A Study of People, Expertise and Sustainable Development.* London: Routledge.

Irwin, A. (2001). Constructing the scientific citizen: Science and democracy in the biosciences. *Public Understanding of Science*, 26(4), 1–18.

Irwin, A., Michael, M. (2003). *Science, Social Theory and Public Knowledge.* Maidenhead, BE: Open University Press.

Irwin, A., Wynne, B. (1996). (Eds.) *Misunderstanding Science? The Public Reconstruction of Science and Technology.* Cambridge: Cambridge University Press.

Israel, G. (2004). Alla scienza non basta la risatina di sufficienza degli zapateri relativisti. *Il Foglio*, 8 ottobre.

Jasanoff, S. (1995). *Science at the Bar: Law, Science, and Technology in America.* Cambridge, MA: Harvard University Press.

Jasanoff, S. (2003). (No) Accounting for expertise. *Science and Public Policy*, 30(3), 157–162.

Jasanoff, S. (2004a) Science and citizenship: A new synergy. *Science and Public Policy*, 31(2), 90–94.

Jasanoff, S. (2004b). (Eds.) *States of Knowledge: The Co-Production of Science and Social Order.* London: Routledge.

Jasanoff, S. (2005). *Designs on Nature. Science and Democracy in Europe and the United States.* Princeton, NJ: Princeton University Press.

Jasanoff, S. et al. (1995). (Eds.) *Handbook of Science and Technology Studies.* Thousand Oaks, CA: Sage.

Jonas, H. (1979). *Das Prinzip Verantwortung.* Frankfurt: Insel.

Jordan, B. (2000). *Les imposteurs de la génetique.* Paris : Editions de Seuil.

Joss, S. (1999). (Eds.) *Public participation in science and technology. Special issue of Science and Public Policy*, XXVI, 290–374.

Joss, S., Bellucci, S. (2002) *Participatory Technology Assessment. European Perspectives.* London: The University of Westminster Press.

Keller, E. F. (1995). *Refiguring Life. Metaphors of Twentieth-Century Biology.* New York, NY: Columbia University Press.

Keller, E. F. (2000). *The Century of the Gene.* Cambridge, MA: Harvard University Press.

Kent, J. (2003). Lay experts and the politics of breast implants. *Public Understanding of Science*, 12(4), 403–421.

Kepplinger, H. M. (1989). *Künstliche Horizonte. Folgen, Darstellung and Akzeptanz von Technik in der Bundesrepublik*. Frankfurt: Campus.

Kolata, G. (1998). *Clone: The Road to Dolly, and the Path Ahead*. New York, NY: William Morrow.

Latour, B. (1991). *Nous n'avons jamais été modernes*. Paris: La Découverte. Translated in English. (1993). *We have never been modern*. New York, NY: Harvester-Wheatsheaf.

Latour, B. (1993). *La clef de Berlin. Petites Leçons de Sociologie de Sciences*. Paris: La Decouverte.

Latour, B. (1997). Socrates' and Callicles' settlement – or, the invention of the impossible body politic. *Configurations*, 5(2), 189–240.

Latour, B. (1999). *Politiques de la nature*. Paris, Editions La Decouverte. Translated in English. (2004). *Politics of Nature. How to Bring the Sciences into Democracy*. Cambridge, MA: Harvard University Press.

Latour, B. (2003). What if we talked politics a little? *Contemporary Political Theory*, 2(2), 143–164.

Latour, B. (2004). Von 'Tatsachen' zu 'Sachverhalten'. Wie sollen die neuen kollektiven Experimente protokollier werden? In Schmidgen, H., Geimer, P. (Eds.) *Kultur im Experiment* (17–36). Berlin: Kultuverlag Kadmos.

Levy-Leblond, J. M. (2003). Scientifiques, encore un effort pour être démocrates! *La Recherche*, 365, Juin, 104.

Lewenstein, B.V. (1992). The Meaning of 'public understanding of science' in the United States after world war II. *Public Understanding of Science*, 1, 45–68.

Lewontin, R. (2000). *It Ain't Necessarily So: The Dream of the Human Genome and Other Illusions*. New York, NY: New York Review of Books.

Liberatore, A., Funtowicz, S. (2003). (Eds.) Democratising expertise – expertising democracy. Special issue of *Science and Public Policy*, 30, 3.

Lippmann, W. (1925). *The Phantom Public*. New York, NY: Harcourt Brace.

MacKenzie, D. (1993). Negotiating arithmetic, constructing proof: The sociology of mathematics and information technology. *Social Studies of Science*, 23, 37–65.

MacKenzie, D. (1996). How do we know the properties of artefacts? Applying the sociology of knowledge to technology. In Fox, R. (Ed.), *Technological Change* (257–263). Reading, MA: Harwood.

MacLeod, C. (1996). Concepts of Invention and the patent controversy in Victorian Britain. In Fox, R. (Ed.), *Technological Change. Methods and Themes in the History of Technology* (137–153). Reading, MA: Harwood.

Malagrida, R. et al. (2004). *An Exhibition to Promote Public Participation in Informed Social Debate on the Use of Embryos*, Resource Document, Paper Presented at the Eighth international PCST conference, Barcelona, 2–5 June, www.pcstnetwork.org

Mannheimer, R. (2003). Via libera ai cibi transgenici? Due italiani su tre dicono no. *Il Corriere della Sera*, 4 agosto.

Maurizi, S. (2004). *Una bomba, dieci storie. Gli scienziati e l'atomica*. Milano: Bruno Mondadori.

Meissner, A. et al. (2005). Generation of nuclear transfer-derived pluripotent ES cells from cloned Cdx2-deficient blastocysts. *Nature*, October 16, Resource Document, www.nature.com.

Meldolesi, A. (2001). *Organismi Geneticamente Modificati. Storia di un dibattito truccato*. Torino: Einaudi.

Melucci, A. (1984). *Altri codici*. Bologna: Il Mulino.

Melucci, A. (1989). *Nomads of the Present. Social Movements and Individual Needs in Contemporary Society*. London: Century Hutchinson.

Merriden, T. (2001). *Irresistible Forces. The Legacy of Napster and the Growth of the Underground Internet*. Oxford: Capstone.

Merton, R. K. (1938a). *Science, Technology and Society in Seventeenth-Century England*. Bruges: St. Catherine Press (Fourth edition, with a new introduction (2001). New York, NY: Howard Fertig).

Merton, R. K. (1938b). *Science and the Social Order*. Reprinted in (1973), *The Sociology of Science*.

Merton, R. K. (1942). *The Normative Structure of Science*. Reprinted in (1973), *The Sociology of Science*.
Merton, R. K. (1957). *Priorities in Scientific Discovery*. Reprinted in (1973), *The Sociology of Science*.
Merton, R. K. (1961). *Singletons and Multiples in Science*. Reprinted in (1973), *The Sociology of Science*.
Merton, R. K. (1963). *Multiple Discoveries as Strategic Research Site*. Reprinted in (1973), *The Sociology of Science*.
Merton, R. K. (1973). *The Sociology of Science. Theoretical and Empirical Investigations*. Chicago, IL: University of Chicago Press.
Meyrowitz, J. (1985). *No Sense of Place. The Impact of Electronic Media on Social behavior*. Oxford: Oxford University Press.
Michael, M. (1992). Lay discourses of science: Science-in-general, science-in-particular, and self. *Science, Technology and Human Values*, 17(3), 313–333.
Michael, M. (1998). Between citizen and consumer: Multiplying the meanings of public understanding of science. *Public Understanding of Science*, 7, 313–327.
Michael, M. (2002). Comprehension, apprehension, prehension: Heterogeneity and the public understanding of science. *Science Technology and Human Values*, 27(3), 357–378.
Mitroff, I. (1974). Norms and counter-norms in a select group of the Apollo moon scientists: A case study of the ambivalence of scientists. *American Sociological Review*, 39, 579–595.
Moore, K. (1995). Organizing integrity: American science and the creation of public interest organizations, 1955–1975. *American Journal of Sociology*, 101, 1592–1627.
MORI-OST. (2005). *Science in Society*. Research conducted for the UK Office of Science and Technology, Department of Trade and Industry, London.
Motluk, A. (2005). Tracing dad online. *New Scientist*, 5 November, 6–7.
Mulkay, M. (1997). *The Embryo Research Debate. Science and the Politics of Reproduction*. Cambridge: Cambridge University Press.
Muller, E. (2005). Sulla nicotina aspro dibattito tra scienziati. *Alfa - Il Sole 24 Ore*, 15 settembre, 16.
Nelkin, D. (1977). Technology and public policy. In Spiegel-Rösing, I., de Solla Price, D. (Eds.), *Science, Technology and Society. A Cross-Disciplinary Perspective* (393–441). London: Sage.
Nelkin, D. (1994). Promotional metaphors and their popular appeal. *Public Understanding of Science*, 3, 25–31.
Nelkin, D., Lindee, S. (1995). *The Dna Mystique. The Gene as a Cultural Icon*. New York, NY: Freeman.
Neresini, F. (2001). Bioetica, medicina e società. In Bucchi, M., Neresini, F. (Eds.), *Sociologia della Salute* (205–237). Roma: Carocci.
New Jersey Statutes Annotated (1991), Declaration of Death, 26: 6A–5.
NISTEP. (2004). Science and Technology Literacy of the Japanese Public, Tokyo: National Institute for Science and Technology, www.nistep.co.jp
Nowotny, H. et al. (2001). *Re-Thinking Science. Knowledge and the Public in an Age of Uncertainty*. Cambridge: Polity Press.
Nyman, D. J., Sprung, C. L. (2000). End-of-life decision making in the intensive care unit. *Intensive Care Med*, 26, 1414–1420.
Observa-Science in Society. (2004). *La crisi delle vocazioni scientifiche*. Rapporto per la Conferenza Nazionale dei Presidi di Scienze, Resource Document, www.observa.it.
Observa-Science in Society. (2005). *Quarto Rapporto su biotecnologie e opinione pubblica in Italia*. Rapporto per il Comitato per la Biosicurezza e le Biotecnologie, Resource Document, www.observa.it.
Observa-Science in Society – Fondazione Giannino Bassetti. (2003). *Terzo rapporto su biotecnologie e opinione pubblica in Italia*. Resource Document, www.observa.it, www.fondazionebassetti.org
OECD (1997). *Promoting Public Understanding of Science*, Paris, Resource Document, www.oecd.org.
Ostellino, P. (2005). Ma la scienza non è teologia. *Il Corriere della Sera*, 4 giugno.

Paccagnella, L. (2000). *La comunicazione al computer*. Bologna: Il Mulino.
Pellegrini, G. (2008). (Ed.) *Technoscientific Innovation. Responsibility and New Models of Democracy in Science in Society Relationship*, Soveria Mannelli: Rubbettino.
Pellizzoni, L. (2005). (Ed.) *La deliberazione pubblica*. Roma: Meltemi.
Peters, H. P. (1995). The interaction of journalists and scientific experts: co-operation and conflict between two professional cultures. *Media Culture & Society*, 17, 31–48.
Peters, H. P. (2002). Scientists as 'Public Experts'. *Tijdschrift voor Wetenschap, Technologie en Samenleving*, 10(2), 39–42.
Phillips, D. M. (1991). Importance of the lay press in the transmission of medical knowledge to the scientific community. *New England Journal of Medicine*, 325, 1180–1183.
Pinch, T., Oudshoorn, N. (2003). (Eds.) *When Users Matter the Co-Constuction of Users and Technologies*. Cambridge, MA: MIT.
Pincock, S. (2005). The chemistry of cash. *Financial Times Magazine*, October 8–9, 10.
Pizzini, F. (1992). *Maternità in laboratorio*. Torino: Rosenberg & Sellier.
Price, F. (1996). Now you see it, now you don't: mediating science and managing uncertainty in reproductive medicine. In Irwin, A., Wynne, B. (Eds.) *Misunderstanding Science? The Public Reconstruction of Science and Technology*. Cambridge: Cambridge University Press.
Price, D. J. de Solla. (1963). *Little Science, Big Science*. New York, NY: Columbia University Press.
Price, D. J. de Solla. (1965). Is technology historically independent of science? A study in statistical historiography. *Technology and Culture*, 6(4), 553–568.
Purdue, D. (1999). Experiments in the governance of biotechnology: A case study of the UK National Consensus Conference. *New Genetics and Society*, 18(1), 79–99.
Randerson, J. (2004). IVF seems safe but only time will tell. *New Scientist*, 30 October, 10–11.
Ravetz, J. R. (1971). *Scientific Knowledge and its Social Problems*. Oxford: Clarendon.
Reich, W. T. (1978). (Ed.) *Encyclopedia of Bioethics*. New York, NY: Free Press.
Revuelta, G. (1998). The New York Times cures cancer. *Quark. Ciencia, Medicina, Comunicación y Cultura*, 12, 48–57.
Riotta, G. (2003). Limiti della scienza, bugie e mezze verità. *Il Corriere della Sera*, 24 Maggio.
Roberts, L. (1991). Fight erupts over DNA fingerprinting. *Science*, 20 December, 721–723.
Rosenberg, N. (1982). *Inside the Black Box: Technology and Economics*. Cambridge: Cambridge University Press.
Rossi, P. (1997). *La nascita della scienza moderna in Europa*. Bari: Laterza.
Rowe, G., Frewer, L. J. (2000). Public participation methods: A framework for evaluation. *Science, Technology and Human Values*, 25(1), 3–29.
Rowe, G., Frewer L. J. (2004). Evaluating public participation exercises: A research agenda. *Science, Technology and Human Values*, 29(4), 512–556.
Rowe, G., Frewer L. J. (2005). A typology of public mechanisms. *Science, Technology and Human Values*, 30(2), 251–290.
Rusconi, G. E. (2002). Laicità e bioetica. *Il Mulino*, 51(4), 668–678.
Rusconi, G. E. (2004). Embrioni prigionieri di Kant *La Stampa*, 28 settembre.
Sartori, G. (1993). *Democrazia. Cosa è*. Milano: Rizzoli.
Scanu, M. (2004). Open archives: Rivoluzione o Conservazione? In *La comunicazione della scienza* (2004), Atti del II Convegno Nazionale di Forlì. Roma: ZadigRoma.
Schiera, P. (1999). *Specchi della politica. Disciplina, melancolia, società nell'Occidente moderno*, Bologna: Il Mulino.
Sclove, R. E. (1998). Better approaches to science policy. *Science*, 279, 1283.
Seagall, A., Roberts, L. W. (1980). A comparative analysis of physician estimates and levels of medical knowledge among patients. *Sociology of Health and Illness*, 2, 317–334.
Shapin, S., Schaffer, S. (1985). *Leviathan and the Air Pump: Hobbes, Boyle and the Experimental Life*. Princeton, NJ: Princeton University Press.
Shaywitz, D., Mellon, D. (2004). How to resolve America's stem cell dilemma. *The Financial Times*, 22 ottobre 2004.
Shiva, V. (2001). *Protect or Plunder? Understanding Intellectual Property Rights*. London: Zed Books.

Snow, C. P. (1960). *Science and Government*. Cambridge, MA: Harvard University Press.
Solomon, S. M., Hackett, E. J. (1996). Setting boundaries between science and Law: Lessons from Daubert v. Merrel Dow Pharmaceuticals, Inc. *Science Technology and Human Values*, 21(2), 131–156.
Sparzani, A. (2003). *Relatività. Quante storie*. Torino: Boringhieri.
Staveloz, V. (2002). *From Science Centre Visitors to Responsible Citizens*, Resource Document, Paper Presented at the Seventh PCST Conference, Capetown, 5–7 December 2002, www.pcst-network.org.
Stehr, N. (2001). *The Fragility of Modern Societies. Knowledge and Risk in the Information Age*. London: Sage.
Stehr, N. (2005). *Knowledge Politics*. Boulder and London: Paradigm.
Tallacchini, M.C. (2002). The Epistemic State: The Legal Regulation of Science, in Mazzoni, C.M. (Ed.) *Ethics and Law in Biological Research* (79–96) (Dordrecht: Kluwer)
Tallacchini, M. C. (2003). Democratizzazione della scienza e brevetti biotecnologici. In Bernasconi, C., Garagna, S., Milano, G., Redi, C. A., Zuccotti, M. (Eds.) *Cellule e genomi* (63–84). Ibis: Pavia.
Tallacchini, M. C. (2005). Scienza e democrazia La scienza destinata a scelte pubbliche. In Guatelli, F. (Ed.), *Scienza e opinione pubblica* (173–218). Firenze: Firenze University Press.
Testa, G. (2006). Che cos'è un clone? Pratiche e significato delle biotecnologie mediche in un mondo globale. In Bucchi, M., Neresini, F. (Eds.) *Cellule e cittadini. Biotecnologie nello spazio pubblico*. Milano: Sironi.
Thompson, J. B. (1995). *The Media and Modernity. A Social Theory of the Media*. Cambridge: Polity Press.
Tiliacos, N. (2004). Sulla vita e sulla morte non si può danzare con leggerezza cinica. *Il Foglio*, 29 settembre 2004.
Touraine, A. (1978). *La voix et le regard*. Paris: Editions de Seuil.
Touraine, A. (1985). An introduction to the study of social movements. *Social Research*, 52, 749–788.
Trench, B. (2008). Internet: Turning science communication inside-out?. In Bucchi, M., Trench, B. (Eds.) *Public Communication of Science and Technology Handbook* (185–198). London: Routledge.
Turkle, S. (1995). *Life on the Screen. Identity in the Age of the Internet*. New York, NY: Simon & Schuster.
Turner, T. (1992). Defiant images: The Kayapo appropriation of video. *Anthropology Today*, 8(6), 5–16.
Turney, J. (1998). *Frankenstein's Footsteps: Science, Genetics, and Popular Culture*. New Haven, CT: Yale University Press.
UK Biotechnology and Biosciences Research Council. (1996). *Ethics, Morality and Crop Biotechnology*. London: UK Dept. of Trade and Industry.
UNCTAD. (2005). *World Investment Report*, Resource Document, www.unctad.org/wir and www.unctad.org/fdistatistics.
van Kolfschooten, F. (2002). Can you believe what you read? *Nature*, 416, 360–363.
Veronesi, U. (2003). Una camera alta per etica e scienza. *Il Corriere della Sera*, 19 maggio.
Viale, R. (2003). Scienza e politica dialogo tra sordi. *La Stampa*, 6 agosto.
von Hippel, E. (2005). *Democratizing Innovation*. Boston, MA: MIT.
Vos, A. (2003). Censura preventiva. *Internazionale*, 28 febbraio, 52.
Wachelder, J. (2003). Democratizing science: Various routes and visions of Dutch science shops. *Science, Technology and Human Values*, 28(2), 244–273.
Ward, B. (2007). The Royal Society and the debate on climate change. In Bauer, M. W., Bucchi, M. (Eds.), *Journalism, Science and Society. Science Communication Between News and Public Relations* (159–172). London: Routledge.
Washburn, J. (2005). *University, Inc. The Corporate Corruption of Higher Education*. New York, NY: Basic Books.

Weber, M. (1922). *Gesammelte Aufsätze zur Wissenschaftslehre*. Tübingen: Mohr. Translated in English. (1949). *The Methodology of the Social Sciences*. New York, NY: Free Press of Glencoe.

Whitley, R. (1985). Knowledge producers and knowledge acquirers. In Shinn, T., Whitley, R. (Eds.), *Expository Science* (3–28). Dordrecht: Reidel.

Wildawsky, A. (1979). *Speaking Truth to Power*. Boston, MA: Little Brown.

Wilkie, T. (1993). *Perilous Knowledge. The Human Genome Project and its Implications*. London: Faber & Faber.

Wynne, B. (1989). Sheepfarming after Chernobyl: A case study in communicating scientific information. *Environment Magazine*, 31(2), 10–39.

Wynne, B. (1995). Public understanding of science. In Jasanoff, S. et al. (Eds.) *Science Technology and Society Handbook* (361–389). Thousand Oaks, CA: Sage.

Wynne, B. (1996). May the sheep safely graze? A reflexive view of the expert-lay knowledge divide. In Lash, S., Szerzinsky, B., Wynne, B. (Eds.) *Risk, Environment and Modernity* (44–83). London: Sage.

Wynne, B. (2001). Expert discourses of risk and ethics of genetically manipulated organisms: The weaving of public alienation. *Politeia*, 17(62), numero monografico su *Politica della scienza e diritto. Il rapporto tra istituzioni, esperti e pubblico*, 51–75.

Yearley, S. (1992). Green ambivalence about science. Legal-rational authority and the scientific legitimation of a social movement. *British Journal of Sociology*, 43, 511–32.

Yearley, S. (1995). The environmental challenge to science studies. In Jasanoff, S. et al. (Eds.), *Science Technology and Society Handbook* (361–89). Thousand Oaks, CA: Sage.

Ziman, J. (2000). *Real Science. What it is, and what it means*. Cambridge: Cambridge University Press.

GPSR Compliance

The European Union's (EU) General Product Safety Regulation (GPSR) is a set of rules that requires consumer products to be safe and our obligations to ensure this.

If you have any concerns about our products, you can contact us on ProductSafety@springernature.com

In case Publisher is established outside the EU, the EU authorized representative is:

Springer Nature Customer Service Center GmbH
Europaplatz 3
69115 Heidelberg, Germany

Batch number: 10837190

Printed by Printforce, the Netherlands